Igneous Petro gy

Contents

Acknowledgements

We are grateful for permission to reproduce the following illustrations:
Fig. 1.1, U.S. Geological Survey; Fig. 1.6a, R.T. Holcomb, U.S. Geological Survey; Fig. 1. 6b, J.D. Griggs, U.S. Geological Survey; Fig. 2.1 Professor P.E. Baker; Fig. 2.2, The Institute of Geological Sciences; Fig. 3.3 The Joint Matriculation Board; Fig. 3.11, Dr J.E. Treagus.

The following figures have been based on illustrations from the sources indicated: Figs 1.2, 1.3, 1.5, Thomas, A.J. *Geology Teaching Vol 6 No 2*; Figs 1.4, 3.4, 3.5, 3.6, *Internal Processes*, Open University; Figs 3.19, 3.24, 5.3, Read, H. and Watson, J. *Introduction to Geology, Vol. 1.*

Foreword

This book covers the general nature and distribution of igneous activity, the classification of igneous rocks and the evidence which the rocks themselves provide about their origins. The main emphasis of the book is on intrusive igneous activity; further information on surface vulcanicity should be sought in other texts, many of which contain abundant illustrations of modern volcanoes.

Except where otherwise acknowledged, the photographs in this Unit have been taken by the authors. The scale bar on such photographs is 5 cm in length.

There are various exercises built into the text. You should try to complete these, indicated by italic type, before going any further.

Units of measurement. The derived S.I. unit of pressure is the pascal (Pa):
1 Pa = 1 newton per square metre
1 Kbar = 10^8 Pa

1
Modern Volcanic Eruptions

Fig. 1.1 The eruption of Mount St Helens, May 18 1980.

Introduction

In March 1961 a little-known event took place in the remote South Atlantic. A ship had unexpectedly become stuck in the open ocean! What was the cause of this potentially dangerous episode? An underwater volcano had erupted on the sea-bed beneath and pumice rose to float on the surface. Pumice completely surrounded the ship and rendered her quite helpless by clogging the inlets to the water cooling system for the ship's engines. It was quite a while before the efforts of seamen with long poles combined with the freshening wind to free the vessel and enable her to resume normal duties.

Thousands of miles away, in 1980, another natural event occurred which was to horrify the world. Mount St Helens erupted. This volcanic peak in the western USA had begun to give notice of its intention when minor earth tremors started in March 1980. Minor eruptions followed and the authorities took carefully planned measures to minimise any possible damage by evacuating people from the probable danger zone. Nonetheless, when the main eruption occurred on 18 May, the effects were devastating (Fig. 1.1).

A large bulge which had developed on the

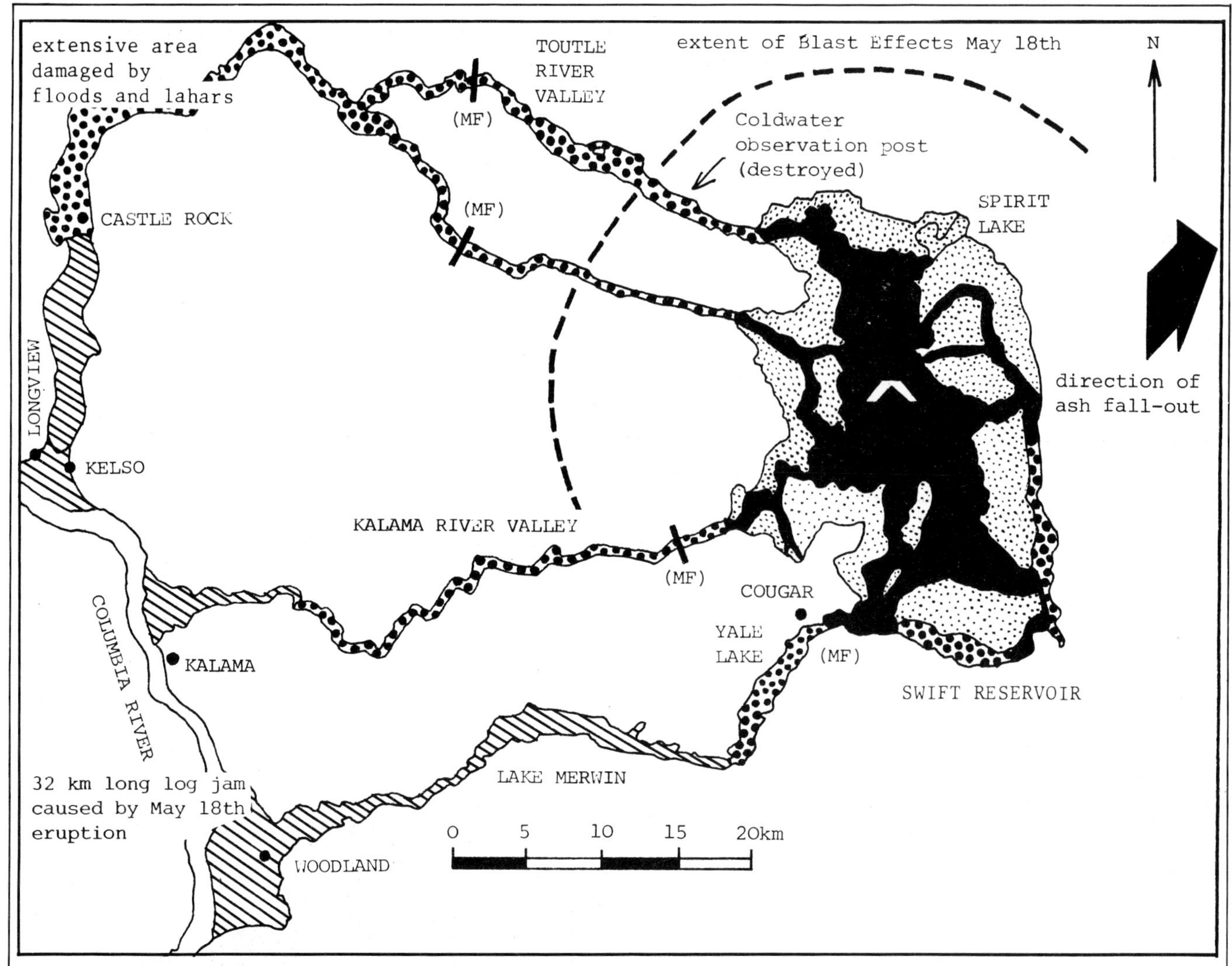

ZONES OF POTENTIAL HAZARDS AROUND MOUNT ST HELENS

pyroclastic surges, lava flows, mud flows and floods

lahars (mud flows) and flows

(MF) approximate limit of thick lahars (mud flows) of May 18

ash clouds associated with pyroclastic surges

floods

dam

Fig. 1.2 The effects of the eruption of Mount St Helens, May 18 1980.

flank of the volcano collapsed, producing a massive landslide. Within seconds, the gas-filled magma which had been 'bottled-up' beneath was blasted into tiny fragments and dust particles, which rolled down the mountainside as an incandescent cloud, travelling at speeds of up to 180 km h^{-1}. A few minutes later, an ash 'fountain' shot up some 19 km into the sky, destined to fall out afterwards over a vast area, covering homes, shops and streets with a thick grey blanket. Three days later the ash cloud reached the Atlantic coast over 4000 km away and in another 17 days it had completely circled the globe at an altitude of 10 km.

Two hours after the eruption, snow and ice from the summit, melted by the heat and mixed with the ash, formed mudflows known as lahars. These careered down neighbouring valleys at speeds of 80 km h^{-1}. Only the driver of a fast car on a clear highway would have been able to escape such forces (Fig. 1.2). Virtually no lava was emitted from Mount St Helens.

The eruption was estimated to have had the power of 10 megatons of TNT and it left 61 people dead or missing and caused an estimated £1,000,000,000 damage to property. This was in spite of the precautions taken to clear the area, precautions which, before the event, were considered by many to be unduly strict.

Yet, it has been estimated that the eruption was only 2% as powerful as the 1815 eruption

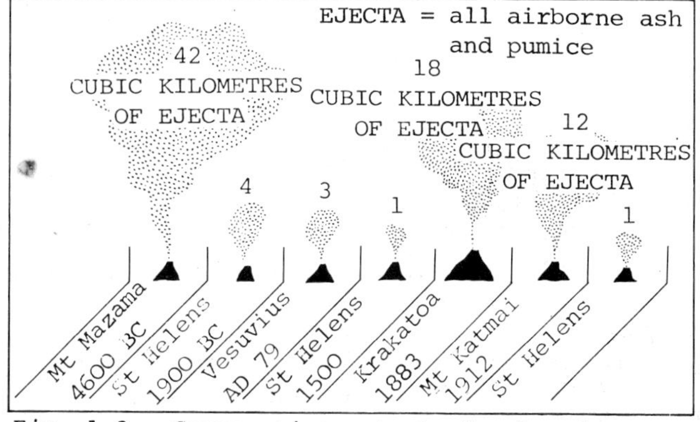

Fig. 1.3. Comparative output of ash and pumice from several volcanoes.

of Tambora in the East Indies. Many other volcanoes have produced far more airborne ash and pumice than Mount St Helens in 1980, as

Fig. 1.3 shows. Even the volcano itself produced four times as much material in 1900 BC than it did in 1980 AD.

Why should one volcanic eruption pass unnoticed by all except a few sailors and another give rise to the declaration of an emergency in the United States?

Why do some volcanoes quietly produce millions of cubic metres of lava with scarcely any ash, whilst others violently expel equally huge quantities of ash and only negligible amounts of lava?

In many cases, volcanoes erupt in different ways at different times of their existence. Why is this? Whilst a volcano is forming on the earth's surface, what is happening at depth beneath it? Can we find evidence of former deep-seated igneous activity, now exposed to view by uplift and erosion? If so, can we establish any link with the magmas which we can still observe in active volcanoes?

These and a host of other questions form some of the problems which igneous petrologists set out to solve. In endeavouring to answer some of them we shall range across quite a wide spectrum of ideas, all of which aid our understanding of the igneous rocks and the processes which created them.

Magmas and their Characteristics

The eruption that stranded the ship at sea took place underwater and apart from collecting pumice, there were few further observations which could be made. However, this type of eruption is typical of many volcanoes of the ocean floor, some of which have built themselves up above sea level to form volcanic islands. Volcanoes of this quieter type are mostly found on oceanic ridges such as the Mid-Atlantic Ridge. The island of Tristan da Cunha, which lies adjacent to the Ridge, provides us with a good example. Its eruptive history is similar to the underwater activity which affected the ship.

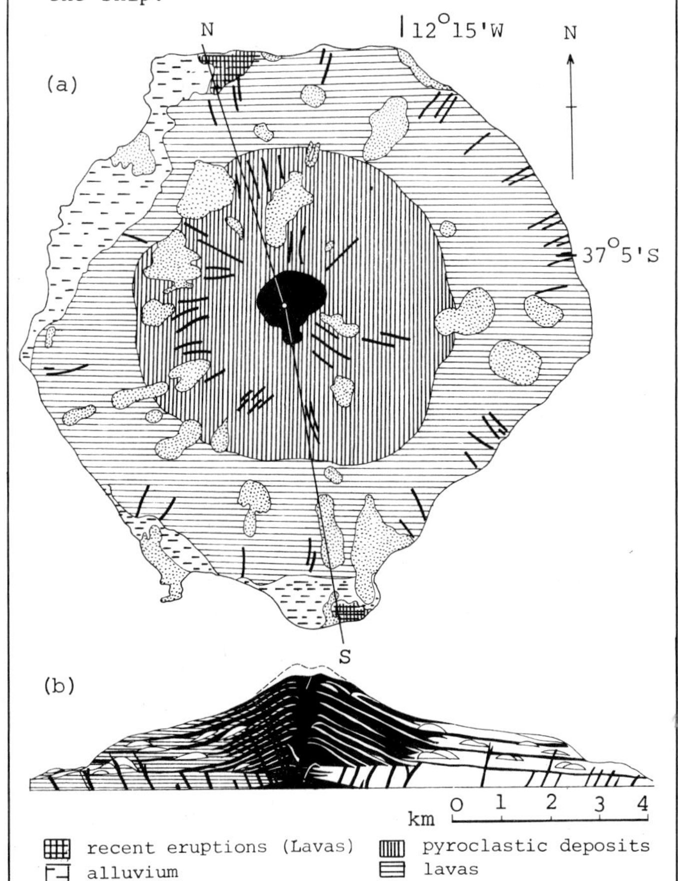

Fig. 1.4 (a) Simplified map of Tristan da Cunha showing various features. (b) Diagrammatic cross-section of the Tristan da Cunha central-vent volcano.

Tristan da Cunha erupted unexpectedly in 1961 and its population had to be hastily evacuated. There was no loss of life, since the main product of the eruption was a thick flow of mobile lava, although this unfortunately obliterated the crawfish canning factory which was the main source of income. It also filled up the only bay on the island where boats could land and extended the coastline a further 500 m. After the main eruption, a team of scientists landed on Tristan da Cunha and studied it as intensively as some of their ill-fated colleagues were to do nearly 20 years later on Mount St Helens.

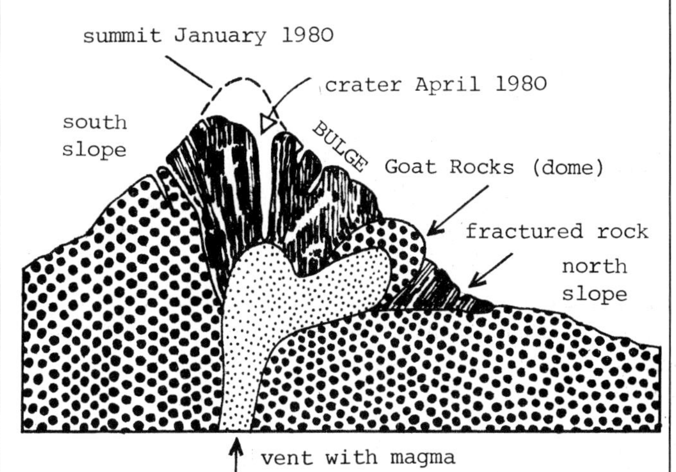

Fig. 1.5 Diagrammatic section of Mount St Helens shortly before the main eruption of May 18 1980.

Careful mapping of each volcano has enabled us to draw sections of their likely structure (Figs. 1.4 and 1.5). The Mount St Helens section shows the volcano a few minutes before the huge bulge slid down the mountainside, releasing the pressure on the magma beneath, which was blasted out as gas-laden dust and ash. Most of the previously existing structure of the volcano was composed of congealed magma and debris. None of the old lava had flowed very far, since any which did not explode was exuded into a high-sided dome before it solidified.

By contrast, much of Tristan da Cunha, and presumably the submarine mound on which it rests, is composed of lava, interspersed with

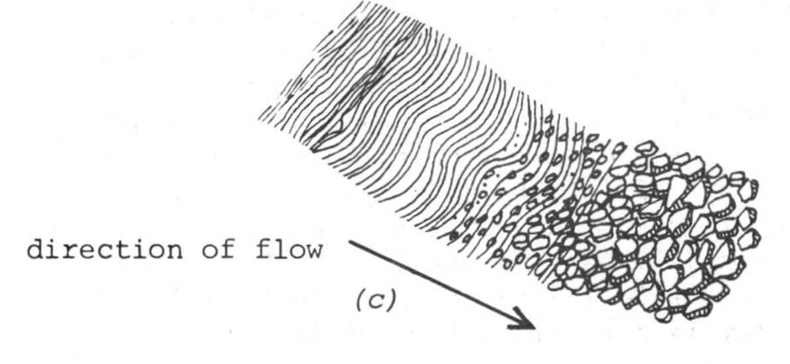

Fig. 1.6 The relationship between the structure of a lava and its position in the lava flow. (a) Ropy lava (pahoehoe), Kilauea, March 1974. (b) Blocky lava (aa), Kilauea, October 1977. (c) Lava with both pahoehoe and aa surfaces

ash bands. Individual flows can be traced for considerable distances and it is evident that the lava was far less viscous (i.e. more mobile) than in the case of Mount St Helens.

In seeking reasons for the differences, a first reaction might be to say that the sea water has a moderating influence on the behaviour of the oceanic volcanoes which is not present on land. However, volcanoes which normally behave in this mild manner are mostly limited to the oceanic ridges. Volcanoes of the island arcs, such as the East Indies, have been known to erupt with even more ferocity than Mount St Helens. Of the volcanoes named in Fig. 1.3, Mount Mazama (now Crater Lake, Oregon) and Mount St Helens are over 130 km inland, Vesuvius and Katmai are within a few kilometres of the coast and Krakatoa is the remains of a volcanic island off the coast of Sumatra. Proximity to water does not seem to provide the answer.

We have already indicated that the behaviour of a magma depends upon its viscosity. Can we determine the factors which affect this property? In the case of the more predictable volcanoes, much can be done by direct, close observation, experiment and sampling of all the products of vulcanicity, ranging from the molten lava itself to the gases which are given off. Temperatures can be measured by thermocouple or optical pyrometer. Observatories such as that of the US Geological Survey on Hawaii have long contributed a great deal to our knowledge. Collecting similar information from a volcano as unpredictable as Mount St Helens is more hazardous and indeed one geologist was killed whilst trying to observe the eruption in 1980. Nonetheless, many valuable data have been gathered, and further experiments have been done in the laboratory on remelted lavas and artificial melts.

The results of such work show that there are three main factors which control the viscosity and hence the likely eruptive properties of magmas. The first of these is temperature. This is illustrated in Fig. 1.6 which shows two markedly different types of lava flow on Kilauea, in Hawaii. The lava in Fig. 1.6a has a 'ropy' surface, known to the Hawaiians as pahoehoe, whilst Fig. 1.6b shows a crumbly, blocky lava flow, known locally as aa. It might be thought that the two are quite unrelated, but there are many cases where pahoehoe passes downstream into aa without any significant differences in the chemical composition of the two forms (Fig. 1.6c). The more viscous downstream part of the flow would clearly have been at a lower temperature, than the mobile upstream part. Other factors, however, such as the internal turbulence of the flow must also be important, or pahoehoe would never solidify but would always change to aa as it cooled.

Such differences in temperature at different stages of a single lava flow are reflected on a bigger scale when one compares the temperature at which an ocean ridge lava erupts with that of the more explosive lavas, when they reach the surface. The former is normally between 1050°C and 1200°C whilst the latter may be as low as 700°C to 800°C.

Another factor is the gas content of the magma. Considerable quantities of gases are produced during volcanic activity, notably water vapour, carbon dioxide, sulphur dioxide and trioxide, nitrogen, carbon monoxide, hydrogen, argon, sulphur and chlorine. Most of these gases are held in solution in the magma until it erupts, whereupon the release of pressure allows them to escape into the atmosphere. Experiment has shown that a rise in dissolved gas content lowers the viscosity and renders a lava more mobile.

The third main factor is the silica content of the magmas. The vast majority of igneous rocks are comprised mainly of silicate minerals, i.e. minerals which contain the elements silicon and oxygen as part of their chemical structure. Some rocks contain 'free' silica in the form of quartz but the biggest part of most rocks is composed of silica in combination with other elements such as calcium, sodium, potassium, aluminium, iron and magnesium. Whatever the actual mineral content, it is not unduly difficult to crush a lava sample and measure the proportions of each chemical component. It is then that we find a distinct correlation between the silica content of the lava and its behaviour when molten. Generally, lavas of high silica content also have a high viscosity; in other words, they are 'sticky' and will not readily flow.

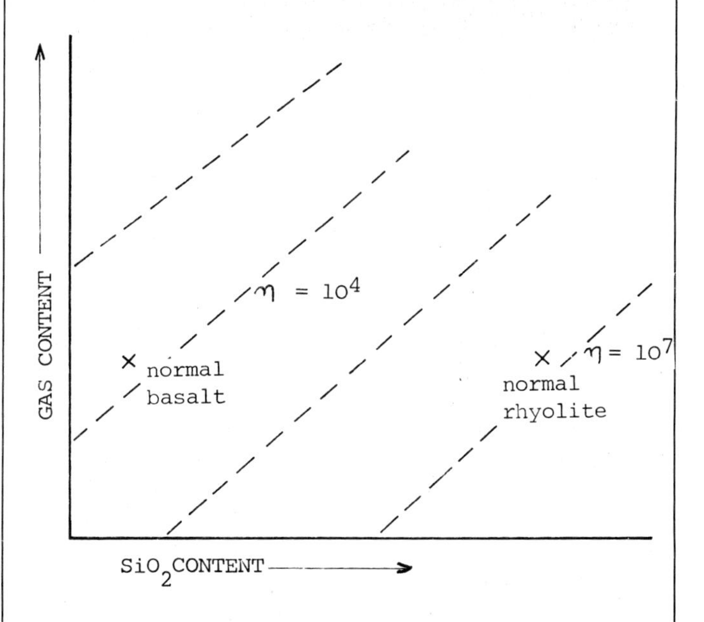

Fig. 1.7 The relation of silicate melt viscosity to SiO_2 and volatile content.

Figure 1.7 summarises, in semi-quantitative form, the influence of gas content and silica content on magma viscosity. Viscosity (η) is traditionally measured in poises (1 poise = 10^{-5} newton seconds per square metre). The graph shows the normal positions of two extremes of lava type, roughly at their temperature of crystallisation at atmospheric pressure. These are basalt (a mobile lava of viscosity 10^4 poise) and rhyolite (a sticky lava of viscosity 10^7 poise).

2
The Classification of Igneous Rocks

The Need for a Scheme of Classification

Such considerations of viscosity alone may help to explain the differences between volcanoes, but we also need to enquire why there are different magma types in the first place and what the ultimate origin of magmas may be. This involves studying phenomena at greater depth within the earth and considering how other igneous rocks are related to the lavas. Figure 2.1 shows clearly that there often is a connection between magmas

Fig. 2.1 Dyke cutting through lavas and ashes, Montagu Island, South Sandwich Group.

erupted as lavas and those which consolidate below ground. The photograph, taken on an island in a chain of recently active volcanoes in the South Atlantic, shows a sequence of lava and ashes which has been partly eroded away to reveal a wall-like dyke. This is made of igneous rock which is chemically similar to the lavas, but which differs from them in that the crystals had longer to form and are therefore larger.

Figure 2.2 is of part of the Cuillin Hills, Isle of Skye, which are composed of yet another igneous rock of the same composition but of coarser grain size still. No one can observe such intrusive rock being formed. It is believed to have cooled considerably more deeply in the crust than the dyke, so it cannot be so easily linked with surface volcanic activity. There are, however, so many chemical and mineralogical similarities

Fig. 2.2 Coire na Creiche, Cuillin Hills, Skye. The hills in the background are composed of gabbro (coarse-grained). In the foreground is an inclined sheet of medium-grained dolerite cutting the gabbro.

to lavas being erupted today that it is natural to seek some links. Obviously, measurements of magma temperature, gas content and rate of flow are quite impossible once the rock is formed, so we must turn to other criteria. In doing so, we hope to develop a method of classification which, if it is to be of any value, must cover the whole range of igneous rocks, from lavas to deep-seated intrusions, from ancient to modern. Any system must be flexible enough to fulfil a wide range of needs.

Geological investigations may include the examination of hand specimens in the field, detailed study of thin slices of rocks beneath the microscope, chemical analyses of crushed rocks or of extracted minerals, as well as observations of actual volcanic processes as outlined above. Many of the products of igneous activity have economic uses, so here is another need for careful definitions. Students will also be painfully aware of the fact that they are often expected to identify odd lumps of rock straight from the drawer, without reference to any helpful aids such as knowledge of the field relationships, rock chemistry, or even thin sections for the microscope! Inevitably, any one system of classification is bound to be in the nature of a compromise, which will need adaptation by the various specialist interests, but the basic framework set out on the following pages has gained general acceptance among geologists.

Principles of the Classification of Igneous Rocks

Chemical composition

We have already indicated that many of the differences between lavas are due to variations in their chemistry, which provide one possible basis for classification. Although at one time laborious, it is now technically feasible for a well-equipped laboratory to measure, quite speedily, the chemical composition of a crushed sample of igneous rock. Results are usually expressed in terms of the oxides of the various elements present, not because they occur in this form but because this is how they were traditionally processed as part of the operation.

TABLE 2.1 COMPOSITIONS OF SOME IGNEOUS ROCKS (AVERAGE WEIGHT %)

COMPONENT	THOLEIITE BASALT	ALKALI BASALT	SPILITE	ANDESITE	TRACHYTE	RHYOLITE	PERIDOTITE
SiO_2	50.0	46.0	51.0	60.0	63.0	73.0	43.5
Al_2O_3	16.0	15.0	14.0	17.0	18.0	13.0	4.0
Fe_2O_3	2.0	4.0	3.0	2.0	2.5	0.5	2.5
FeO	7.0	8.0	9.0	4.0	1.5	1.5	10.0
MgO	8.0	9.0	4.5	3.5	0.5	0.5	34.0
CaO	12.0	9.0	7.0	7.0	1.0	1.5	3.5
Na_2O	2.5	3.5	5.0	3.5	7.0	4.0	0.5
K_2O	0.5	1.5	1.0	1.5	5.0	4.0	0.3
TiO_2	1.5	3.5	3.5	0.5	0.5	0.5	1.0
H_2O	0.5	0.5	2.0	1.0	1.0	1.5	0.7

Table 2.1 gives the results of analyses of some typical igneous rocks. The importance of the silica content of a lava has already been stressed and it is this which has been chosen as the main basis of chemical classification of all the igneous rocks. Many years ago, arbitrary divisions were chosen and the spectrum of igneous rocks divided into four categories. The names of each of these were based upon a misconception of the actual properties of silica, but the names have remained, in spite of attempts to find better ones. The categories are shown in Table 2.2.

TABLE 2.2

PERCENTAGE OF SiO_2 in BULK CHEMICAL COMPOSITION	NAME OF CATEGORY
	ultrabasic
— 45	basic
— 52	intermediate
— 66	acid

The main advantage of this system is that it enables the geologist to relate the solidified rock to its probable magmatic source. For example, the lavas of oceanic ridges look very similar to those of ocean fracture zones and yet there are differences in their chemistry which reflect their different origins (i.e. tholeiite-basalt and alkali-basalt in Table 2.1). On the other hand the lavas, dyke rock and deep seated intrusive already referred to look very different from each other and yet analyses of them show that they are chemically almost identical.

The main disadvantage is one of inconvenience. We need names to apply to specimens as we find them, without having to await the laboratory's report. Another complication is the rather arbitrary way in which the divisions were drawn up in the first place. After many thousands of analyses, we know that the division at 66% SiO_2 coincides with a minimum in the frequency curve of all the rocks analysed, so it is quite well chosen (Fig. 2.3). The other two dividing lines are badly selected, however, the 52% line actually coinciding with a peak of abundance of rock types!

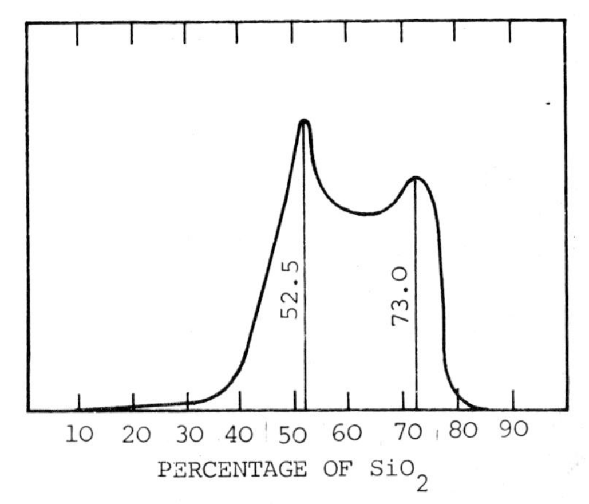

Fig. 2.3 The frequency distribution of silica percentage in analyses of igneous rocks.

In spite of these problems, the concept of a range of rocks from ultra-basic to acid is of considerable value, although in practice the label is more frequently derived from other criteria related only approximately to the silica percentage. Variation in other chemical components is also of importance, yet it is not expressed by this method.

Colour

We have seen that rocks may only be grouped on the basis of their chemistry after laboratory analysis. By contrast, one of the most obvious properties of a rock which one can easily record is its colour. Igneous rocks range from pale grey or white through to black. This often, although not always, reflects significant differences in rock chemistry or mineral content and it may be used as a rough basis for classification. The terms used are based on Greek words and are as follows:

light coloured - leucocratic 0-30% dark minerals
medium coloured - mesocratic 30-60% dark minerals
dark coloured - melanocratic over 60% dark minerals

The method provides a useful basis for rough divisions but it reveals nothing about the genesis of the rock. Also, whilst most of the melanocratic rocks are basic or ultrabasic types there are some acid rocks which are black. A good example is the shiny black volcanic glass, obsidian (Fig. 3.7b).

More recently, other terms have entered common usage in an attempt to describe concisely the dominant types of minerals present in the rock. Thus 'felsic' describes a rock in which feldspar and quartz (silica) are major constituents. 'Mafic' is used for rocks which contain some feldspars but which are rich in the ferromagnesian minerals (i.e. magnesium and iron (Fe) bearing). 'Ultra-mafic' labels a rock which is almost completely made of ferromagnesian minerals alone. The felsic minerals are mostly light-coloured and the mafic ones dark, so for most practical purposes the terms have come to be applied on the basis of the overall colour of the rock, and are often loosely regarded as alternatives to 'leucocratic', 'melanocratic', etc.

Mineral content

Most students have seen and handled good specimens of a range of minerals. In doing so, they have learned that most minerals have clearly defined properties which enable them to be identified. The majority of igneous rocks are composed of minerals which have crystallised tightly packed together. Identification of them is not so easy as it is for the individual mineral specimens but with a little practice it can be done, either in the hand specimen, or with a thin-section of the rock beneath the petrological microscope, or both. The relative abundance of each mineral can also be estimated. Because of the wide range of chemical constituents comprising rocks from acid to ultrabasic, there is an equally diverse number of minerals which may be present. Here then is a potentially useful way of establishing divisions, both in the field and in the laboratory.

Over 3,000 minerals are known, but, fortunately for the petrologist (person who studies rocks), most of these are uncommon! In practice, the majority of igneous rocks can be described in terms of a dozen or so which are usually referred to as the rock-forming minerals. The main groups of minerals found are quartz, the feldspars, the micas, ferromagnesian minerals such as olivines, pyroxenes and amphiboles, and the iron ores, notably magnetite.

The table in Fig. 2.4 has been constructed empirically from the results of many thousands of mineralogical analyses of igneous rocks. It shows the variation in the proportion of each of the main rock-forming minerals and how this is related approximately to the chemical division described before. The ferromagnesian minerals are listed in order of abundance.

The minerals listed are usually referred to as the essential minerals and it can be seen that there is only a handful of such minerals in each category, e.g. basic rocks consist essentially of plagioclase and a pyroxene such as augite. Other minerals are commonly present, but usually in smaller quantities and these are known as accessory minerals. Examples would include sphene, apatite or zircon in acid rocks. In the

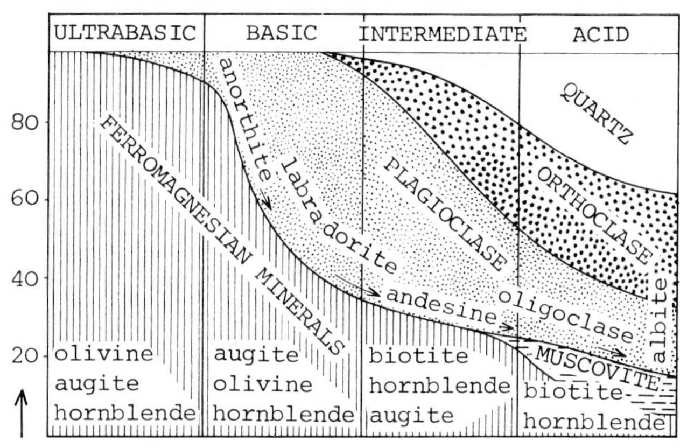

Fig. 2.4 The chief minerals of igneous rocks.

basic category, olivine, although listed in the table, is often regarded as an accessory mineral. When appropriate, its presence is indicated by a hyphen, including the word as a prefix to the rock name, e.g. olivine-basalt.

In identifying igneous rocks from their mineral content, it is usual to approach them systematically:

1. Is quartz present and if so in what proportion? The table shows that acid rocks can contain over 20% quartz, intermediate ones between 20% and a few percent and basic ones very little. When quartz does occur in a basic rock it is usually regarded as an accessory. IT IS VITAL NOT TO CONFUSE THE PERCENTAGE OF THE MINERAL QUARTZ IN THE ROCK WITH THE PERCENTAGE OF SILICA IN ITS BULK CHEMISTRY. They both share the formula SiO_2, yet in the chemical analysis the SiO_2 percentage is the total of all the silica, occurring both in the free state and in combination in other silicate minerals. Thus, a basalt may have a SiO_2 content of 50% and yet have no 'free' quartz.

2. What is the feldspar content and which feldspars are present? The proportion of the rock which is composed of feldspar can usually be estimated in the hand specimen, as can the distinction between potash feldspar (orthoclase etc.) and the plagioclases. The determination of the variety of plagioclase, however, is normally only possible with the aid of a petrological microscope. A slice of rock is ground down until it is so thin that most of its minerals are transparent. This is examined beneath a microscope equipped with polarised light and a rotating stage. The optical properties of minerals in such thin sections are quite characteristic and it is not difficult to identify them to the level required by Fig. 2.4. It is important to be able to work in this detail, since the classification partly depends on the type of plagioclase, ranging from the calcium-rich types (anorthitic) in the basic rocks to the sodium-rich (albitic) in the acid ones. Potash feldspar and the albite variety of plagioclase feldspar are commonly referred to as the alkali feldspars.

3. Are micas present and if so which ones?

The white mica, muscovite, is most often found in the acid rocks and helps to give them their lighter colour. Biotite is often present in acid rocks but is most abundant in the intermediate ones.

4. Which ferromagnesian minerals are present and what are the proportions? The ultra-basic rocks are mainly composed of the darker, denser ferromagnesian minerals and it is perhaps in this category that distinguishing between them is of greatest importance. The table (Fig. 2.4) shows the most common ferromagnesian minerals in each group, and it also shows that some minerals are mutually exclusive. For example, it would be very unusual to find both olivine and quartz in the same rock; if sufficient silica is present to form free quartz, then olivine crystals would not have survived for long in the magma, but would have reacted with the excess silica to form an amphibole or a pyroxene. (See p.20)

It must be evident from the above detail that the mineral content of a rock provides a powerful basis for its classfication. Nonetheless, there are some drawbacks. It does not distinguish between rocks formed at different depths, thus fine-grained rocks often have to await sectioning and microscope examination before a positive identification can be made. It is also easy to ignore possible genetic relationships between rocks by putting them in neat categories. For example, small quantities of acid lavas are found in association with basic ones and may represent later derivations from the same magma.

Texture

By the word texture we mean the grain size of the rock and the relationship between neighbouring mineral constituents. We shall examine some particular textures later but it is sufficient here to note the relationship between the size of the crystals in an igneous rock and the time which it took to cool. The concept will be familiar to anyone who has seen salol crystallise at different rates or has tried to grow large copper sulphate crystals from aqueous solution. Generally, a slower rate of cooling produces bigger crystals.

In geological terms, slower cooling rates are achieved deeper in the crust than at the surface, so the use of grain size, as a criterion for classification, will usually reflect the geological environment in which the rock crystallised. In the past, such principles were probably taken a little too far, and the terms 'volcanic', 'hypabyssal' and 'plutonic' were applied to the hand-specimen to indicate, respectively, whether the rock was produced at the surface, at moderate depths, or deep in the crust. Such ideas are most useful in the field, where larger scale structure may also be observed, but it is a little risky in the hand specimen. For example, not all fine-grained basalts are lavas; basalt quite frequently occurs as dykes, intruding other strata.

A Classification of Igneous Rocks

In practice, the most useful classification system is one which combines the mineral content of a rock with its texture. The approximate relationship between mineral content and chemical composition has already been shown in Fig. 2.4 and Table 2.2. In Table 2.3 which shows an outline classification of igneous rocks, texture has been plotted against the approximate chemical categories only, for the sake of simplicity.

Both Fig. 2.4 and Table 2.3 therefore need to be read together in order to relate mineral content and texture. For example, a basalt is a fine-grained basic rock and gabbro is a coarse-grained one, but they are chemically similar and contain the same essential minerals, namely calcic plagioclase and pyroxene.

Petrologists have set up a host of names for varieties of igneous rocks but in Table 2.3 we have used the rock names in their general sense. For example, the name 'granite' covers a family or clan of coarse-grained acid rocks.

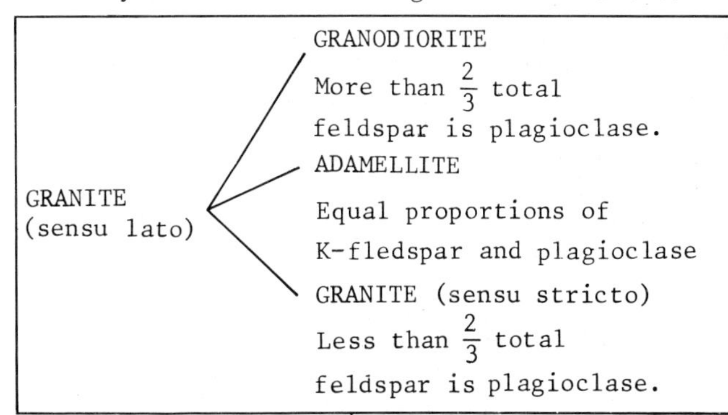

TABLE 2.3 A CLASSIFICATION SCHEME FOR IGNEOUS ROCKS

"ULTRABASIC"	"BASIC"	"INTERMEDIATE" plagioclase> K feldspar	K feldspar> Plagioclase	"ACID"	VOLCANIC & MINOR INTRUSIVE ROCKS
(rare)	basalts	andesites	trachytes	rhyolites obsidians	fine-grained or glassy
(rare)	dolerites	microdiorites	microsyenites	microgranites	medium to fine-grained
peridotites serpentinites	gabbros	diorites	syenites	granites	PLUTONIC ROCKS coarse to medium- grained

The alternative names in the fine-grained acid rock categories reflect marked textural differences, rather than mineralogical ones. The intermediate rocks are divided into two groups because of significant differences in mineralogy, which have long been recognised by the use of these names. In the trachytes and syenites, the proportion of alkali feldspars (Na and K rich) is relatively high. The andesites and diorites are proportionately richer in the calcium feldspars.

It must be stressed again that such a fitting of rocks into pigeonholes is a convenient but rather artificial way of seeking order in the natural world. Many geologists prefer alternative systems, but all would agree that whatever system is chosen, rocks do not fall into neat slots but form parts of a continuously varying spectrum.

Since so much of our understanding of igneous rocks depends upon microscope examination of thin sections, a series of drawings of such sections follows (Fig. 2.5). The drawings are in the same order as in Table 2.3, for ease of reference. The minerals have been shown by partly stylised symbols, which approximate more closely to the view seen in plane polarised light than to crossed polars (i.e. two pieces of Polaroid with their polarisation axes at right angles).

Fig. 2.5 A selection of igneous rocks in this section.

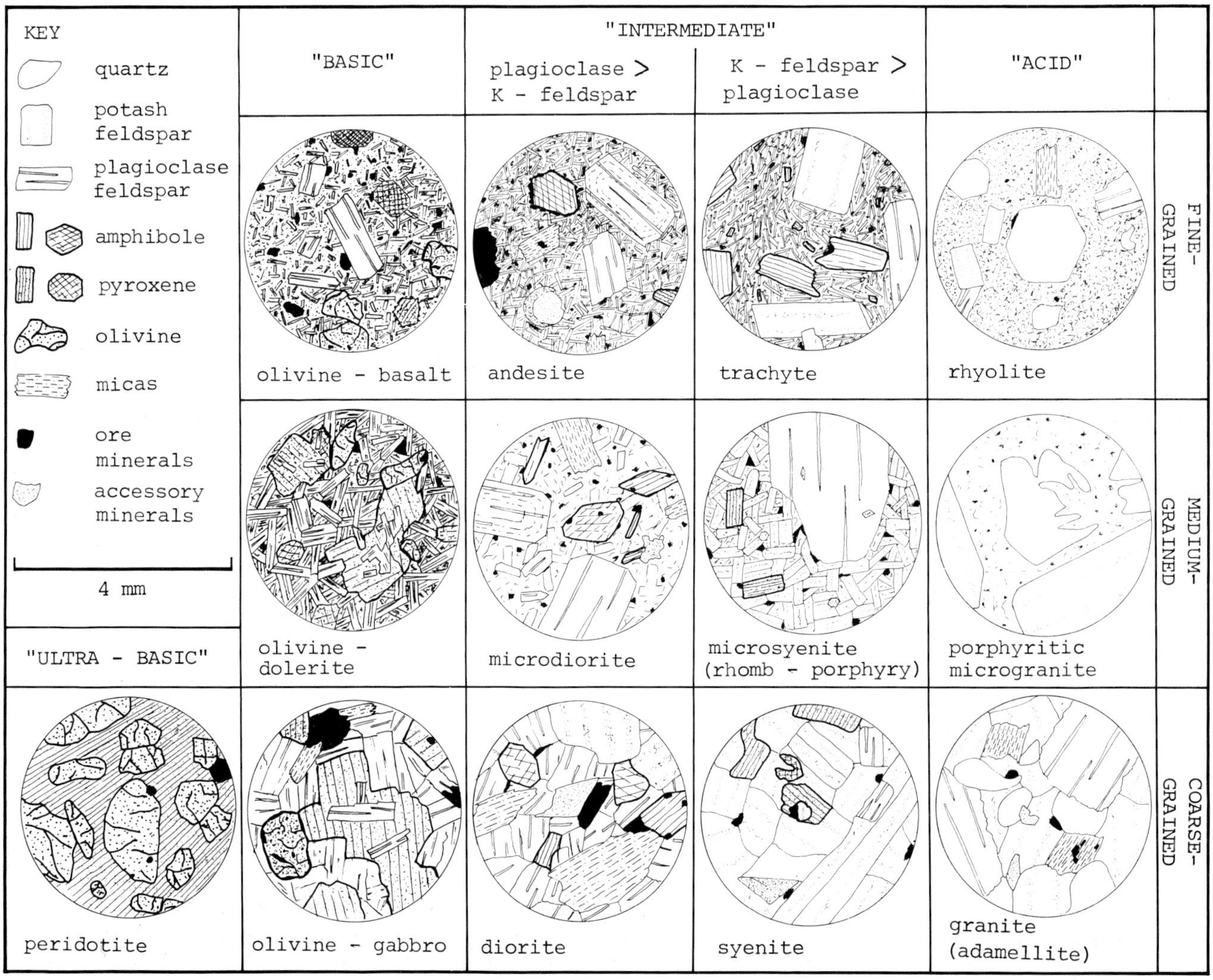

KEY: quartz; potash feldspar; plagioclase feldspar; amphibole; pyroxene; olivine; micas; ore minerals; accessory minerals; 4 mm	"BASIC"	"INTERMEDIATE" plagioclase > K - feldspar	"INTERMEDIATE" K - feldspar > plagioclase	"ACID"	
	olivine - basalt	andesite	trachyte	rhyolite	FINE-GRAINED
"ULTRA - BASIC"	olivine - dolerite	microdiorite	microsyenite (rhomb - porphyry)	porphyritic microgranite	MEDIUM-GRAINED
peridotite	olivine - gabbro	diorite	syenite	granite (adamellite)	COARSE-GRAINED

3

Determining the History of an Igneous Rock

We have discussed the classification of igneous rocks at some length, since more is at stake than simply assigning a name to a given piece of rock. The use of the correct name tends to make most geologists think not just of a lump of rock, but also of its origin and its significance in global tectonic terms. However, there is more information to be considered which can add a great deal to our knowledge of the origin of an igneous rock. Broadly, this falls into three categories: laboratory work on silicate melts; detailed examination of rock textures; observations of the structure and field relationships of igneous rock masses.

Silicate Melts in the Laboratory

Laboratory experiments have been carried out over the last 70 years or so, since the pioneer work of N L Bowen and his associates in the USA. In view of the very high temperatures and pressures involved, such work requires highly specialised equipment and skills, but it is now possible to imitate natural processes by remelting existing rocks, or by making synthetic ones. We can thus gain vital information regarding the temperatures and pressures at which rocks may begin to melt to produce magmas. It is also possible to predict the cooling history of a magma, which may then be compared with the field and laboratory evidence from the rocks themselves. Some of the most important results concern the effects on melting points of mixing several components or of introducing water vapour into the system.

Figure 2.4 showing mineral content indicates that most igneous rocks consist of two or more essential minerals. Of course, accessory minerals are also usually present. Many experiments have been conducted with more than two components present, but the results are rather difficult to understand in graphic form and we shall limit ourselves to melts with only two components. A simple analogy of the principles involved is the way in which salt lowers the freezing point of water. We shall consider three examples.

The diopside-anorthite system

Diopside and anorthite are two relatively simple silicate minerals. Diopside ($CaMg Si_2O_6$) is a pyroxene and anorthite ($CaAl_2 Si_2O_8$) is an end-member of the plagioclase feldspars. In a way, this mixture may be regarded as a simple version of a basic rock such as basalt. The graph (Fig. 3.1) is derived from the experimental work carried out at atmospheric pressure, and it shows the effect which each component has on the other. Thus, pure diopside melts at 1391°C and pure anorthite at 1660°C. The outcome of mixing the two is to lower the melting point, the amount depending on the proportions of each component present. The lowest melting point on the graph is at 1270°C, corresponding to a composition of 58% diopside and 42% anorthite. This point is known as the eutectic.

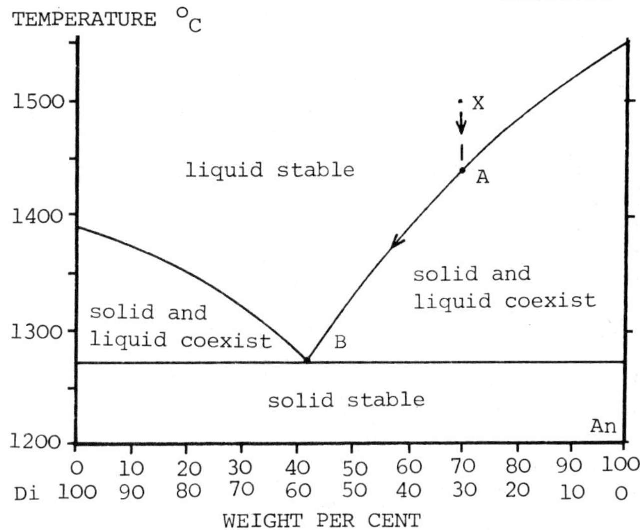

Fig. 3.1 Equilibrium diagram of the system anorthite (An) diopside (Di).

The graph may be used to determine the crystallisation history of a melt as follows. Assume a melt of composition 30% diopside, 70% anorthite at a temperature of 1500°C (point X on the graph). As the temperature falls, there is at first no change in the composition of the melt, nor do any crystals form, since only liquid is stable above the line. When the temperature reaches 1440°C, anorthite begins to crystallise, but not diopside. Because the components of anorthite are being removed from the liquid by crystallisation, the composition of the liquid will change along the line AB. Once the temperature has fallen to 1270°C and the composition has reached the eutectic proportions (Di_{58} An_{42}) then diopside will crystallise. After this, the two minerals crystallise together in the eutectic proportions, and the temperature remains the same until the liquid is all used up and the mass has set solid.

Reheating the solid material is the exact reverse of the above. The first liquid appears at 1270°C and the temperature does not increase until all the diopside and most of the anorthite has melted.

The graph relates to a 'dry' system. The addition of water vapour, which is very often present in real melts, lowers the melting points and moves the eutectic to the right. For example, at a water vapour pressure of 10^9 Pa (10 kilobars), the melting point of pure diopside drops to 1280°C and that of anorthite to about 1130°C. The eutectic is then at 1020°C. The applications of such studies to igneous petrology are quite far-reaching and some will be dealt with in later sections.

Crystallisation of the olivines

The pale green mineral which most of us know as olivine is in fact one member of a whole family of such minerals. The olivines are silicates of iron (Fe^{2+}) and/or magnesium (Mg^{2+}), e.g. $MgFeSiO_4$. The ions of magnesium and iron are each of valency 2 and their dimensions are comparable. (The ionic radii are: Fe^{2+} = 0.83Å; Mg^{2+} = 0.78Å.) This means that Mg^{2+} and Fe^{2+} are easily interchangeable within the well-ordered lattice structure of the mineral and so a continuous range of composition is possible, from an all-magnesium variety of olivine (Mg_2SiO_4 - forsterite) to an all-iron variety (Fe_2SiO_4 - fayalite). This phenomenon of interchangeability of ions is known as isomorphous substitution (isomorphous means 'same shape', i.e. there is no change in the crystal lattice structure throughout the series). The series is also known as a 'solid solution series', since it appears as though one component is 'dissolved' in the other on the molecular scale. Although the chemistry of the olivines is rather simpler than that of the diopside-anorthite system, the melting-point curve is perhaps a little more difficult to follow.

The diagram (Fig. 3.2) shows the cooling curves for the olivine family from forsterite to fayalite. The melting point of pure forsterite is 1890°C and that of pure fayalite is 1205°C.

Consider a magma of composition 50% forsterite, 50% fayalite (i.e. Fo_{50} Fa_{50}) at a temperature of 1800°C, shown as point X. The magma will cool to 1650°C (Y) before any crystals appear. When they do, they will have the composition shown by the lower line at that temperature, i.e. point Z on the graph; a composition of Fo_{80} Fa_{20}. Such crystals will be in equilibrium with the melt. As the temperature drops, however, the composition of the melt changes along the upper line and that of newly-formed crystals changes along the lower line, in the direction of the arrows. This has some interesting implications. Unless the early-formed crystals have been removed from the melt, perhaps by sinking to the bottom of the magma chamber, they will find themselves surrounded by a liquid with which they are no longer in equilibrium. Hence, fresh mineral growth around the margins of such crystals will be more iron-rich than the original core of the crystal and it will appear 'zoned'. In a rock

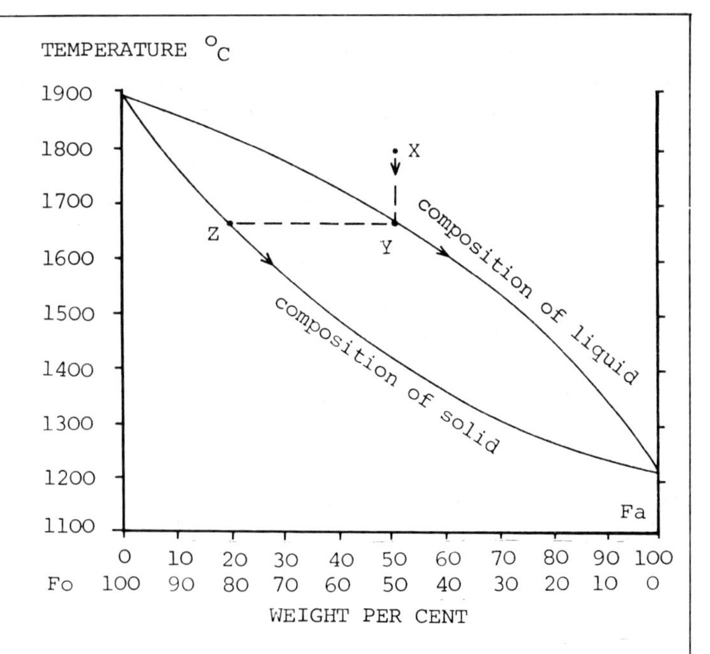

Fig. 3.2 Equilibrium of the system forsterite (Fo)-fayalite (Fa).

which contains larger, early-formed crystals (phenocrysts) set in a finer-grained groundmass, the large crystals may be markedly richer in magnesium than the later, smaller ones. If cooling is slow enough, the whole crystal may be altered by such reaction.

Crystallisation of the plagioclase feldspars

The plagioclase feldspars also form a continuous series of isomorphous substitution (solid solution) between albite ($NaAlSi_3O_8$) and anorthite ($CaAl_2Si_2O_8$).

The crystallisation diagram is shown in Fig. 3.3 as an exercise for you. The answers are given at the end of the Unit. Again, the influence of water vapour (not shown) serves to lower the melting points of the whole system.

Crystallisation of magmas

Real magmas usually contain more than the two components considered in each of the above studies; their crystallisation curves become progressively more complex and they are not dealt with in detail here. Of vital interest, however, is the behaviour of real magmas at different depths in the earth, and much of the experimental work has been directed at studying the effects of pressure as well as temperature. Many magmas also contain appreciable quantities of water vapour and this has a marked influence on their crystallisation history, as already indicated for the diopside-anorthite system.

Basic magmas

Figure 3.4 shows a simplified melting-point curve for basic magma, where temperature is plotted against pressure. The pressure can also be roughly equated to depth of burial in the earth. The curve is that for 'dry' basic magma. Although much water vapour is exhaled during volcanic eruptions, it forms a relatively minor constituent of basic

Study the graph below which shows the crystallisation of plagioclase feldspar under equilibrium conditions. A liquid melt of mixed composition cools to temperature X (1500 °C) at which crystallisation begins. The crystals which form at this temperature may be represented by point Y on the lower curve and have a composition Z (almost pure anorthite). As the temperature continues to fall, the composition of the liquid melt changes in accordance with the upper curve on the graph. The composition of the crystals also changes - both for those already formed, and for those that will form at the lower temperature. The liquid melt thus remains in equilibrium with its crystals.
Figure A, which refers to part (c) of the question, shows a zoned crystal of plagioclase feldspar in which the composition of each layer has *not* changed during crystallisation.
Answer the following questions.

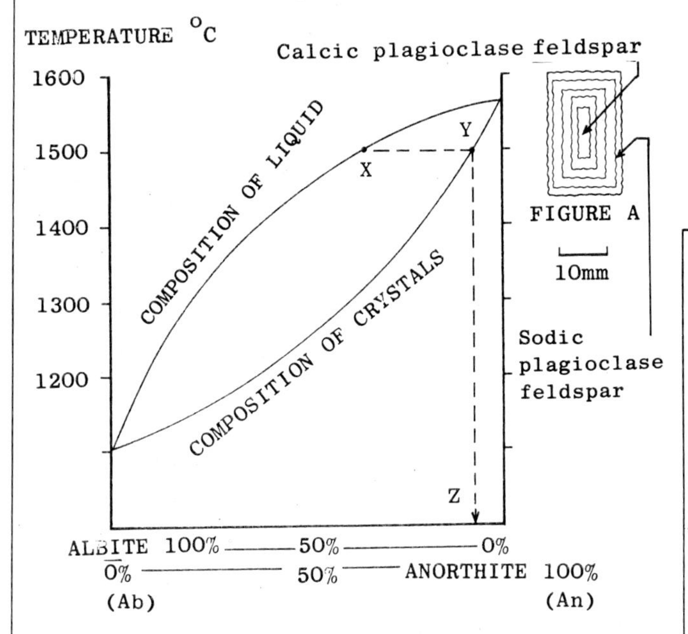

(a) Account for the change in composition of plagioclase feldspar which occurs during crystallisation from a liquid of composition X under equilibrium conditions (very slow cooling).

(b) What are the properties that enable ions to substitute for one another to permit the change from anorthite ($CaAl_2Si_2O_8$) to albite ($NaAlSi_3O_8$)?

(c) In some basic igneous rocks, zoned plagioclase crystals are formed, for example the crystal in Figure A. Explain, briefly, how the zoned crystal could have been formed.

Fig. 3.3 Exercise based on JMB A level paper.

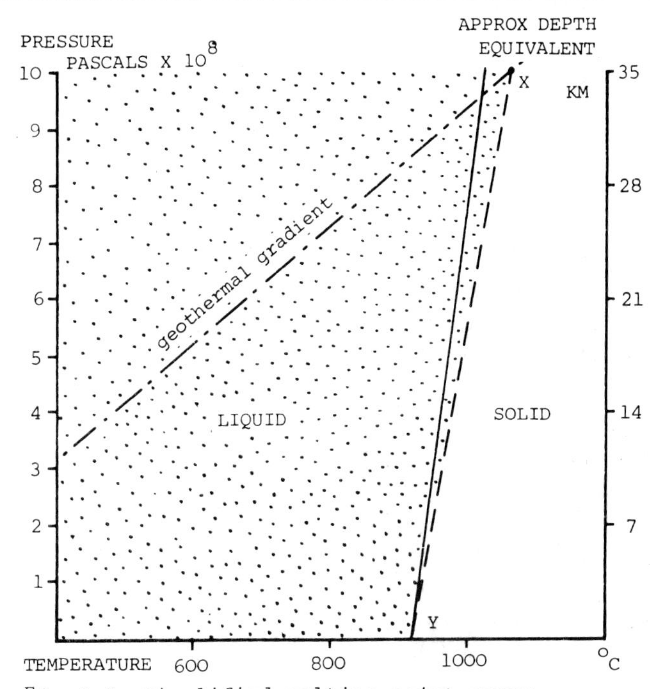

Fig 3.5 Simplified melting point curve for "dry" acid magma.

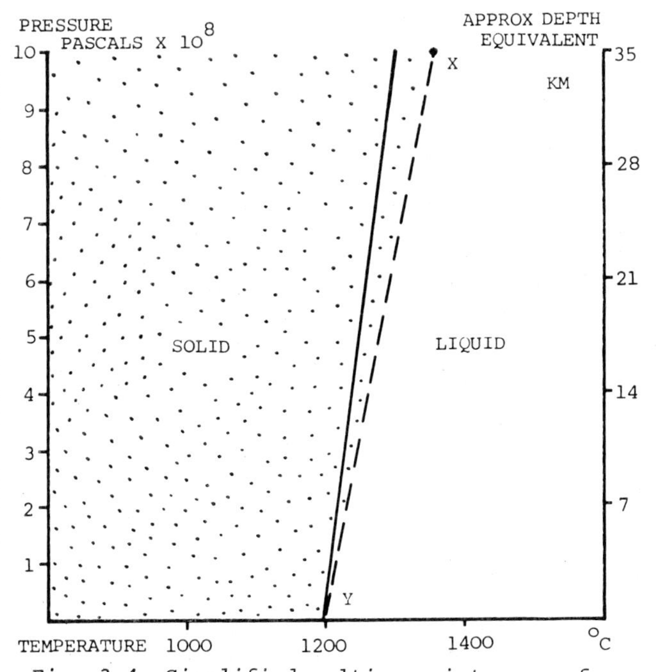

Fig. 3.4 Simplified melting point curve for "dry" basic magma.

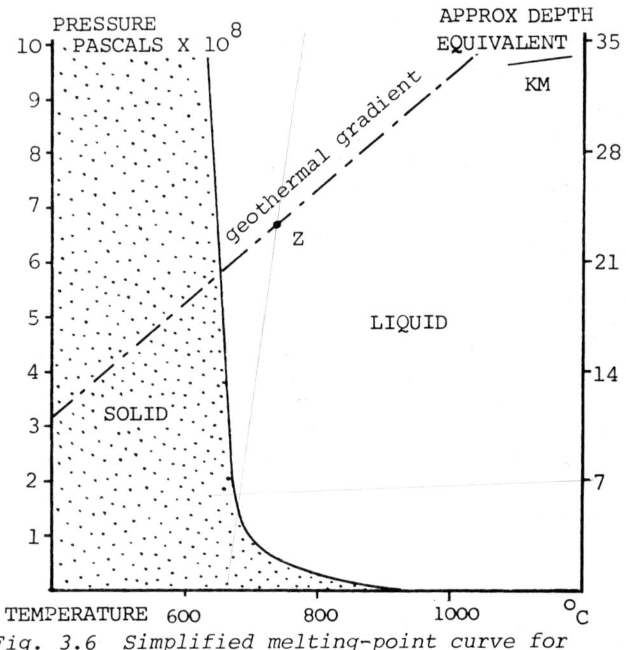

Fig. 3.6 Simplified melting-point curve for "wet" acid magma where water vapour pressure equals load pressure.

magmas and can be ignored for the present purposes. Assume for the sake of this exercise that a basic magma forms at 1350°C and 10^9 Pa pressure. As it rises through the earth, assume that it cools along the line XY (remember that rising in the earth means moving down on the graph). The line XY intersects the melting point curve at the earth's surface, indicating that the magma will still just be in liquid form and capable of erupting as lava.

Although this presents an oversimplified view, it is not too far from the truth and helps to explain why basaltic lavas are so common.

Acid magmas

Figures 3.5 and 3.6 illustrate melting point curves for acid magmas. Figure 3.5 is for 'dry' magmas, whilst Fig. 3.6 is one of many which could be drawn for an acid magma containing water vapour. In this case, the water vapour pressure is assumed to be equal to the load pressure from the overlying crust. The difference between the two graphs is remarkable, the effect of the water vapour being to lower the melting point by as much as 400°C.

Many acid magmas originate by the melting, or partial melting, of the rocks of the lower crust, consequently the 'normal' geothermal gradient within the crust has been added to the diagrams.

It can be seen that for a melt originating at point X, rising in the crust and cooling at the rate of say 4°C per km (XY) it is capable of reaching the surface and erupting as lava.

Now try the same exercise for the 'wet' acid magma, assuming a starting point of Z, and the same rate of cooling. How far will it rise before it begins to crystallise? (See Appendix.)

In practice, it would seem that 'wet' acid magmas are more common than dry ones, perhaps because of the pore-water contained in the crustal rocks from which they are formed. Certainly, acid lavas (rhyolite, obsidian etc) are less voluminous than granite masses. We shall later consider the evidence for the cooling of both basic and acid magmas with some case studies. First, we shall examine the importance of the textures of igneous rocks as an aid to understanding their origin.

Textures of Igneous Rocks

General textures

A considerable amount may be learned about a rock from its texture. The importance of grain size in classification and as an approximate guide to the depth at which the rock crystallised, has already been outlined. However, there exists a wide variety of detailed textures, many of which have frightening Greek names and some of which are not easily understood. Some textures are clearly visible to the naked eye in the hand specimen, some require the assistance of a good hand-lens, and many can only be studied in thin section beneath the microscope. At the risk of divorcing rock textures from the rest of the data about the rock, it may be helpful at this stage to catalogue the various possibilities, with some indication about their origins. In some cases, a direct link will be observed with the crystallisation diagrams of the last section.

Table 3.1 outlines the main characteristics of each texture but the drawings or photographs are shown separately (Fig. 3.7).

You should try to match each drawing or photograph to the appropriate description. Inevitably, the drawings show textures taken out of context. When you have checked answers (at the end of the Unit) it would be a useful exercise to turn back to the thin-section drawings in Fig. 2.5 and see how many textures can be identified in the whole rock and what can be inferred about its origin.

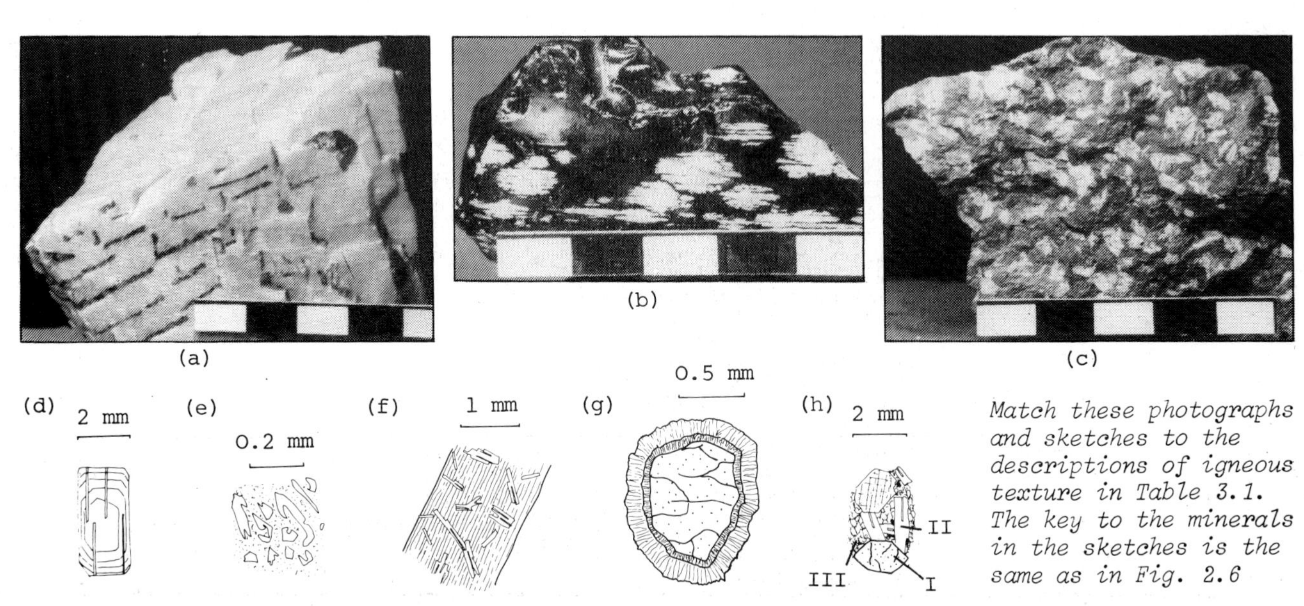

Match these photographs and sketches to the descriptions of igneous texture in Table 3.1. The key to the minerals in the sketches is the same as in Fig. 2.6

Fig. 3.7 Textures of igneous rocks in the hand specimen and in thin section.

TABLE 3.1 TEXTURES IN IGNEOUS ROCKS

TEXTURES	EXAMPLES	DESCRIPTION	A POSSIBLE INTERPRETATION	
Glassy texture	Devitrified obsidian	Usually black and shiny in hand specimen, with conchoidal fracture. Ancient volcanic glasses often have tiny crystals arranged radially in 'spherules', or along curving 'perlitic' cracks.	Result from supercooling of magmas where cooling was too rapid for crystallisation and the magma set as a glass. With time the glass crystallises, or devitrifies into tiny crystals.	
Grain size (not illustrated in Fig 3.7)	Coarse grained	Crystals of the ground mass are easily seen and mostly identifiable with the naked eye.	Crystals had plenty of time to grow around a limited number of nucleii. Usually typical of plutonic rocks.	COOLING TIME several millions of years
	Medium grained	Crystals of the ground mass seen with tne naked eye but hand lens needed for identification.	Crystals formed more quickly around a greater number of nucleii. Usually typical of minor intrusions.	↓
	Fine grained	Crystals of the ground mass not distinguishable with the naked eye and not identifiable with a hand lens.	Crystals formed very quickly around many nucleii. Typical of lavas and chilled margins of minor intrusions	perhaps less than a year
Grain shape	Euhedral	Well-formed crystals showing perfect or near-perfect crystal form.	Usually the first crystals to form in the magma and therefore unrestricted.	
	Subhedral	Grains show an imperfect but still recognisable crystal form.	Formed at a time intermediate between the early and late stages of crystallisation.	
	Anhedral	Grains show no regular crystal form.	Usually the last crystals to form, filling up gaps. Many rocks consist largely of equidimensional anhedral or subhedral crystals. Their texture is referred to as granular.	
Intergrowth textures	Granophyric	The rock consists of quartz and feldspar grown together in curious embayed crystal shapes. Associated with medium to fine-grained rocks.	Probably results from crystallisation at or near the eutectic point.	
	Graphic	Another quartz-feldspar intergrowth found in coarse or very coarse-grained rocks. Quartz crystals grow in angular sheets looking like ancient cuneiform writing. Ratio of quartz to feldspar usually 30 : 70	*Try to decide for yourself.**	
Reaction textures	Zoned crystals	Common in plagioclase feldspars. Crystals show concentric 'rings', the central portion generally being more calcium-rich than the outer zones which are relatively sodium-rich.	Reaction has taken place between crystal and liquid in a continuous solid solution series under relatively quick cooling. There is not time for the whole of the early-formed crystal to react with the changed composition of the melt.	
	Corona structure	This appears as a type of zoning, except that the zones are composed of different minerals, e.g. olivine in the core, with a rim of pyroxene, followed by an outer rim of amphibole.	Also results from reaction between early-formed crystals and the melt as its composition changes, so that the two are no longer in equilibrium. Occurs in minerals forming discontinuous reaction series, i.e. different minerals develop rather than variations of the same mineral.	
Some other textures	Ophitic texture (a type of poikilitic texture)	Small euhedral plagioclase crystals are enclosed by one large augite crystal. Most common in dolerites. Other pairs of minerals with similar relationships may occur. The general name for such pairs is poikilitic texture.	*Try to decide for yourself.**	
	Porphyritic texture	Large crystals, usually euhedral forms, are set in finer-grained groundmass. The large crystals are termed phenocrysts. Common in lavas and in some granites (e.g. Shap Granite).	*Assuming that crystal size is mostly controlled by rate of crystallisation, try to make your own decision about this one too.* *	

* *Answers to the illustrations and to the problems marked * will be found at the end of the unit.*

Textures limited to particular groups of rocks

Pegmatites and aplites

Quite often, limited portions of plutonic rock masses and offshoot dykes exhibit unusually coarse-grained and unusually fine-grained rocks. The former are known as pegmatites (Fig. 3.8a) and they sometimes contain crystals of a metre or more across. A giant crystal of spodumene nearly 15m long was recorded in South Dakota. The most common types are associated with granite plutons and consist largely of quartz, feldspars and micas. However, they may also contain minerals rich in former volatiles such as boron and fluorine. Tourmaline is a good example.

The finer grained rocks are known as aplites and usually have a sugary appearance (Fig. 3.8b). They may contain similar minerals to pegmatites.

Both groups of rocks are regarded as the products of late-stage magmatic activity when the residual magmas were rich in water vapour and volatiles. The relationships between pegmatites and aplites remain rather speculative (Fig. 3.8c).

Lava textures

Lavas exhibit many of the textures listed in the table, but they are also characterised by other textural features which are not normally associated with intrusive rocks. The main ones are the following:

Vesicular texture

Vesicles are cavities in the lava produced by the escape of gases (Fig. 3.9a). Vesicular lavas are most typical of the tops of lava flows, where there is no pressure of overlying material to keep the gas in solution. Sometimes a vesicular lava may occur near the base of a flow; in this case it has been carried down from the cooled top surface as the inside of the flow moved forward, rather like a tank-track.

Amygdaloidal texture

This Greek-looking term is derived from the word for almonds. It describes a lava where

(a)

(b)

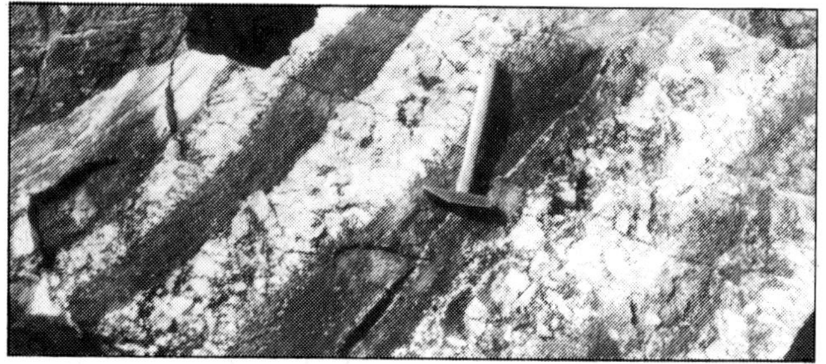

(c)

Fig. 3.8 (a) A pegmatite with tourmaline, Porthleven, Cornwall. (b) A narrow aplite vein in granite, Shap Fell Cumbria. (c) Aplite and pegmatite veins in close association, Porthleven, Cornwall.

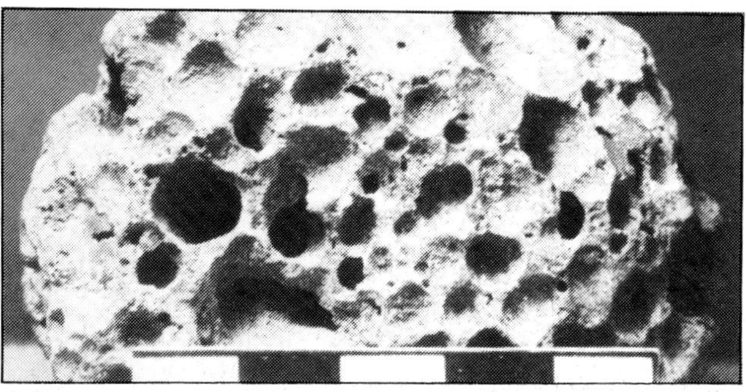

(a)

(b)

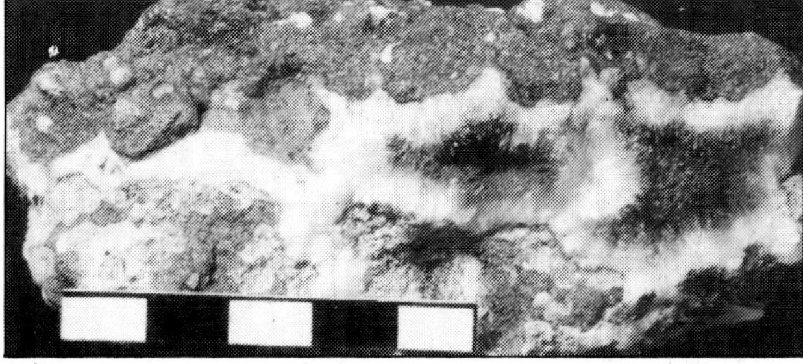

(c)

Fig. 3.9 (a) Vesicular lava, Mediterranean. (b) Amygdaloidal lava with calcite filling almond-shaped vesicle. Locality unknown. (c) Amygdaloidal lava with two generations of zeolite crystals filling elongated vesicles. Locality unknown.

the vesicles have been filled in by later generations of crystals. In a weathered lava, these sometimes stand out as oval shapes, looking rather like sugared almonds (Fig. 3.9 b and c).

Flow-banded textures

Some of the more viscous lavas, usually those of acid or intermediate composition, exhibit a rough banding of their constituents (Fig. 3.10a). The bands are usually continuous for several centimetres and are frequently highly convoluted. A special case is trachytic texture, which is typical of trachytes and consists of a 'swirling' of tiny laths of potash feldspar crystals (Fig. 2.5).

(a)

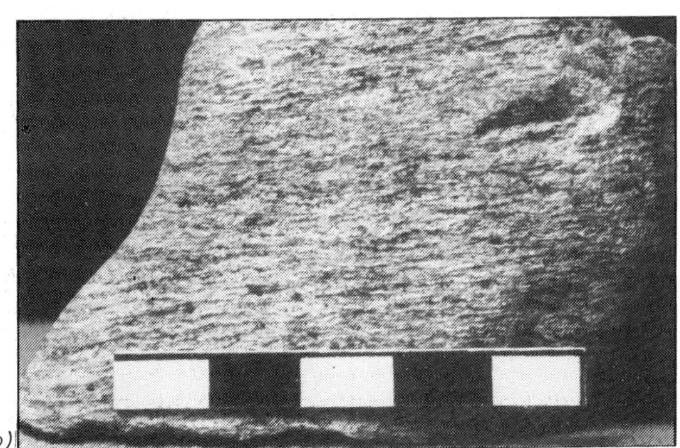
(b)

(c)

Fig. 3.10 (a) Flow-banded rhyolite, Snowdon. (b) Ignimbrite, locality unknown. (c) Bedded tuff, Eifel Mountains, West Germany.

Textures of pyroclastic rocks

True flow-banding of viscous lavas is often confused with the texture of an ignimbrite, which is not produced by a lava flow at all. Ignimbrites (sometimes called welded tuffs) originate by the deposition and rapid 'welding' together of dust particles which travel down the slopes of a volcano in a great incandescent cloud, or nuée ardente. The hot dust cloud originates from the shattering of very viscous lava which has generally formed a dome-like mass within the surface layers of the volcano. The characteristic features of an ignimbrite are the discontinuous nature of the bands and the presence of crushed shards of glass looking like the letter 'Y' on its side (Fig. 3.10b).

The mode of origin of an ignimbrite is often referred to as a 'pyroclast flow' to distinguish it from the other types of pyroclastic activity, where the heat is not sufficient to produce the welded texture. These other pyroclastic rocks show great variation in particle size, ranging from badly sorted agglomerates through a variety of tuffs (Fig. 3.10c) to fine volcanic dust, which may settle to form a type of loess. The agglomerate in Fig. 3.11 is unusual in that it exhibits graded bedding, the fragments probably having fallen into water.

Fig. 3.11 Agglomerate showing graded bedding.

In conclusion, comment on the texture of the rock in the photograph (Fig. 3.12). See the back of this Unit for an answer.

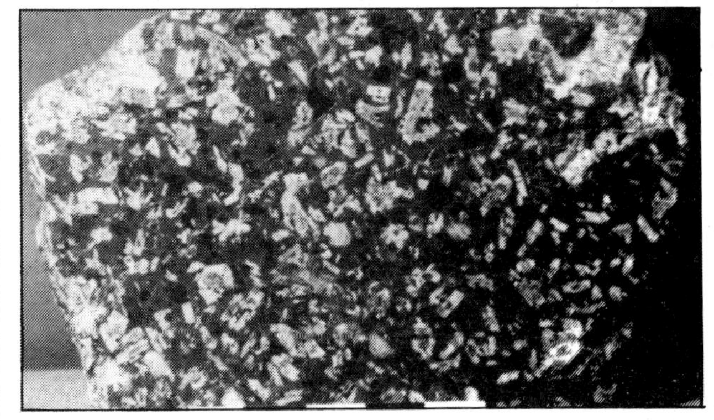

Fig. 3.12 Comment on the texture of this rock.

Major Basic Intrusions

Much can be learnt about the way magma behaves during crystallisation and about its origins by studying major intrusions. Most of the major intrusions fall conveniently into two broad groups, the basic and the granitic, using the terms in their broadest sense, and these will be considered separately.

There are a number of very large basic intrusions which can be quite informative about the way in which magma crystallises. In dealing with major basic intrusions it should not be forgotten that the rate of cooling is completely different from extrusive events and some major intrusions can take thousands or even millions of years to finally consolidate.

We have already outlined the significance of laboratory work by N L Bowen and others. In a sense, a large basic intrusion may be thought of as a gigantic natural laboratory where the results of processes acting over a long period of time may be preserved. Observations of reaction rims (Fig. 3.7) and the order of crystallisation from silicate melts (see earlier) have led to the formulation of an ideal sequence in which minerals could be expected to crystallise. This sequence has become known as 'Bowen's Reaction Series' and it consists of two lines of descent for

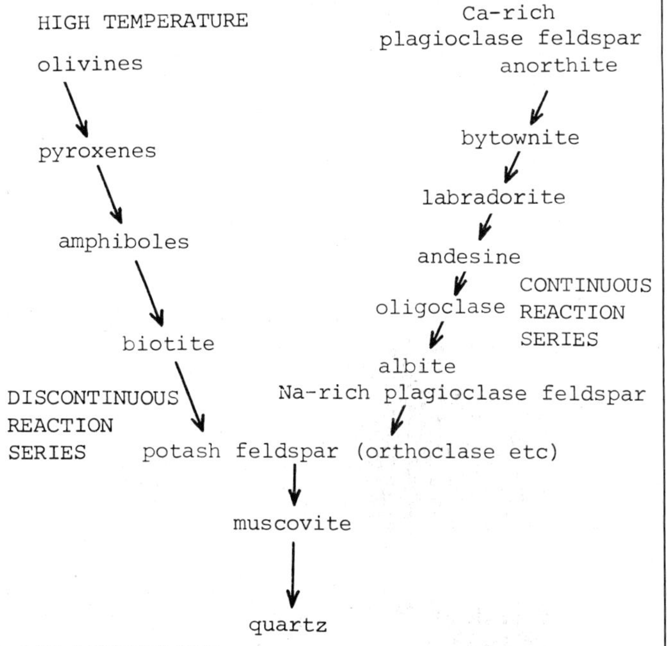

Fig. 3.13 Bowen's Reaction Series.

silicate minerals (Fig. 3.13). One comprises the ferro-magnesian minerals which form a discontinuous reaction series giving rise to the kind of reaction rims illustrated in Fig. 3.7. The other consists of the plagioclase feldspars, which form a continuous reaction series, sometimes seen in zoned crystals. The two series are related to temperatures of crystallisation and unite at their lower temperature ends. The minerals at the high temperature ends of the series (the olivines and calcium-rich plagioclase feldspars) have the highest melting points of their respective series and would therefore be expected to crystallise first as a basic magma cools. They are also the densest minerals of their series and therefore tend to sink within the melt. These would be followed by the pyroxenes and then the amphiboles with the most acidic minerals such as muscovite and quartz crystallising last.

It is obviously not possible to watch an igneous intrusion crystallising, so one has to use indirect evidence to determine its cooling history. Much of this comes from studies of detailed features such as the zoning of crystals which can be related to the experimental work on magmas already described. There are also a few intrusions which show a systematic distribution of minerals within them which can be related to Bowen's Reaction Series.

The Palisades Sill

The Palisades Sill of New Jersey, USA, provides us with such an example. This sill is over 300 metres thick and it is intruded into Triassic sandstones (Fig. 3.14).

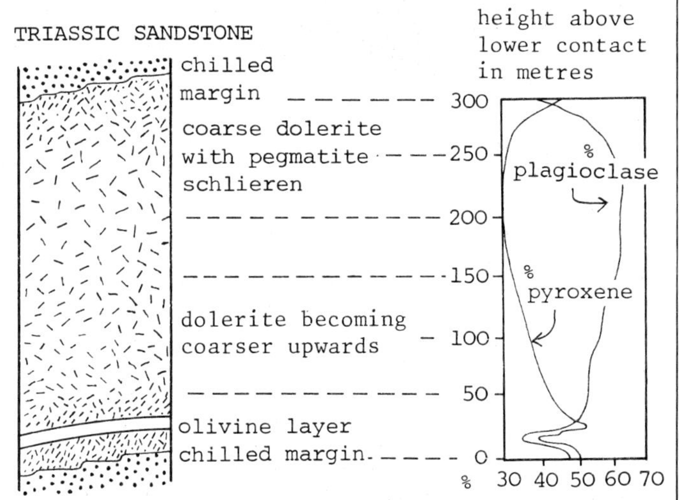

Fig. 3.14 Section through Palisades Sill.
Fig. 3.15 Variation in percentages (by weight) of plagioclase and pyroxenes.

Initially, chilled margins were formed at the top and bottom of the intrusion, each about 16 m thick, and the composition of these rapidly chilled portions of the intrusion is thought to be the same as that of the original magma. This shows it to have been a basic magma with tholeiitic affinities. As the rest of the magma took longer to cool than this chilled margin, it is coarser grained and is thus dolerite rather than the basalt seen in the chilled margin. The dolerite becomes coarser upwards.

As the main part of the intrusion started to cool the silicate minerals crystallised out in the same order as shown in Bowen's Reaction Series. As can be observed from this sequence the olivine and calcium-rich plagioclase feldspars will tend to crystallise first. However, the normal pattern of events whereby these minerals would be expected to react with later liquids has not been followed in the Palisades Sill and a curious distribution of minerals within the intrusion has resulted.

Above the lower chilled margin is a layer 10 m thick which contains 25% olivine - the so-called olivine layer. These olivine crystals have a high specific gravity relative to that of the magma so that after formation they settled out towards the bottom of the intrusion. Some of the crystals may have taken as long as 200-300 years to sink down through the intrusion. Minerals which sink and accumulate at the bottom of intrusions are known as cumulate minerals. Their sinking is one way in which an originally homogeneous magma may begin to crystallise in various fractions, each of which may be significantly different from another. The processes by which such differences are created are known collectively as differentiation.

This trend of differentiation continues throughout the intrusion and although it does not lead to any spectacular layers like the olivine layer, changes can be detected in the composition and proportions of the pyroxenes and plagioclase feldspars. The feldspars become more sodium-rich whilst the pyroxenes increase in iron content in the higher parts of the intrusion. The proportion of the plagioclase feldspars also increases higher in the intrusion (Fig. 3.15). Table 3.2 indicates the contrasts in rock chemistry between different parts of the intrusion.

Because of the early settling out of the more basic components of the intrusion the residual liquid becomes more 'acidic' so that the last part of the intrusion to crystallise, which is about 40 metres from the top, consists of acidic lenses or schlieren containing quartz and orthoclase feldspar. These have crystallised from aqueous solutions which were concentrated in the last part of the intrusion after the crystallisation of the anhydrous phase. They tend to be coarse-grained, which probably reflects the concentration of water and other volatiles as the liquid phase of the magma was gradually displaced upwards by the sinking crystals.

It is worth noting that recent studies have shown that this explanation may be something of an oversimplification. In fact there are 16 metres of normal dolerite between the top of the chilled margin and the olivine layer. This has led to speculation that there may have been more than one pulse of magma intruded, with the olivine settling out at the base of the second pulse.

TABLE 3.2 CHEMICAL COMPOSITIONS OF DOLERITES FROM PALISADES SILL (Weight %)

CONSTITUENT	1	2	3
SiO_2	51.91	48.28	52.32
TiO_2	1.25	0.82	0.97
Al_2O_3	15.31	9.36	16.54
Fe_2O_3	0.98	2.14	1.58
FeO	9.31	11.54	8.66
MnO	0.08	0.12	0.12
MgO	7.52	17.48	5.43
CaO	9.71	7.00	9.68
Na_2O	2.30	1.59	2.32
K_2O	0.79	0.41	1.03
H_2O+	0.93	0.99	0.84
H_2O-	0.15	0.06	0.40
P_2O_5	0.18	0.11	0.06
CO_2			0.16
	100.42	99.90	100.11

1 Average undifferentiated dolerite from chilled margin
2 Olivine dolerite from olivine rich layer
3 Quartz dolerite 150 m above lower contact

The presence of these minor lenses of acid rock in the Palisades Sill is predictable from Bowens Reaction Series but we need to examine an example of a larger intrusion to see if significant quantities of acid rocks can be generated in this manner.

The Skaergaard intrusion

The Skaergaard Intrusion of east Greenland is one of a series of Tertiary intrusions and it is an excellent example of a so-called layered basic intrusion. It has the shape of an inverted cone and, since it has been closely studied by many workers, it is very well documented, although still not perfectly understood.

Around the edges of the intrusion there is a fine grained olivine-gabbro series known as the marginal border group (Fig. 3.16), which is believed to represent the chilled margin

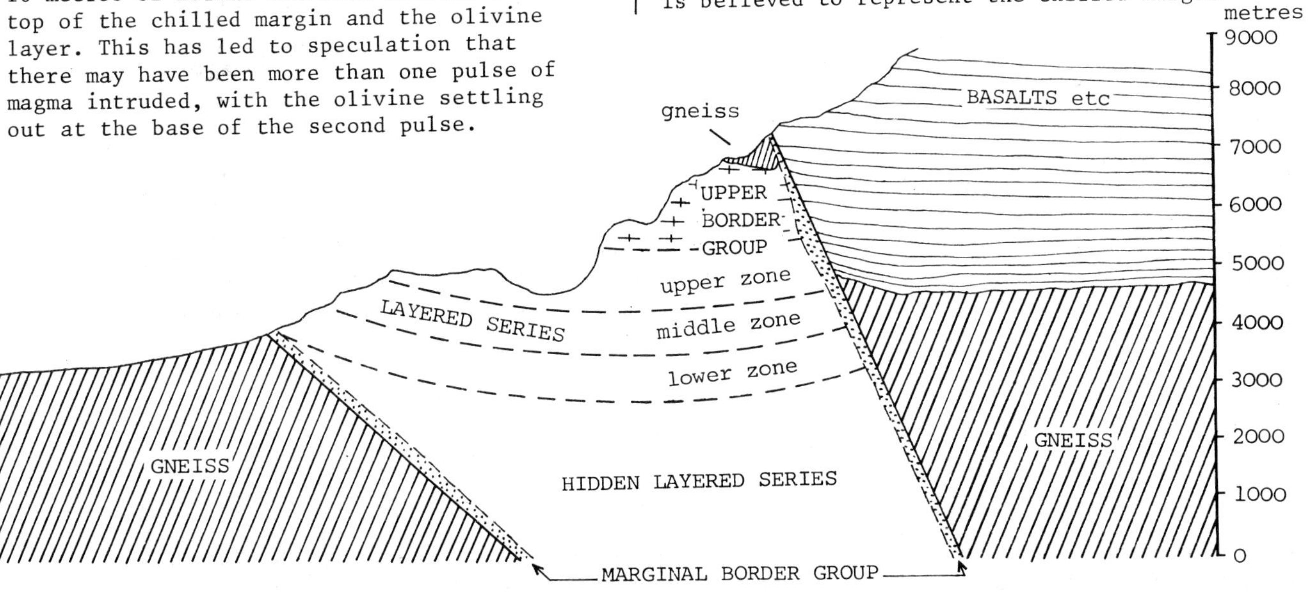

Fig. 3.16 Section through the Skaergaard Intrusion.

of the intrusion, and so, just as with the chilled margin of the Palisades Sill, its composition is thought to be the same as that of the original magma. The marginal border group has suffered contamination due to melting of the surrounding rocks, in particular where it is in contact with gneiss, which has a lower melting point than the Tertiary basalt lavas. However, sufficient samples which are considered to be uncontaminated have been obtained to state with some confidence that the original magma was tholeiitic in character.

The main part of the intrusion is made up of the layered series. As the name suggests this is composed of a succession of layers, each of which is turned up at the edges (Fig. 3.16) and the series represents cooling from the base upwards. By no means all of the layered series is exposed as can be seen from the diagram. Indeed, it has been estimated that over 70% of the intrusion is not exposed and therefore some of the deductions and conclusions reached are, of necessity, rather tenuous.

Within each of the layers, a clear stratification is seen with the dark, heavy minerals, pyroxene and olivine, at the base and the less dense, lighter coloured plagioclase feldspar concentrated near the top of each layer.

This layering is thought to have an origin very much like that of graded bedding in sedimentary rocks. As convection currents swept down through the intrusion they transported crystals with them and deposited them as a 'mush' on the floor of the intrusion. Gravitational settling of the heavy, dark minerals within these cumulate layers would then be responsible for the light and dark bands. Within the layered series, troughs can be seen which have been scoured out by the convection currents.

Less obvious, but of greater significance, is the trend of changing mineral compositions throughout the layered series. Passing upwards, the minerals tend towards the low temperature end members of their respective solid solution series. Thus the ferromagnesian minerals such as olivine and pyroxene become enriched in iron and the plagioclase feldspar becomes richer in sodium. The complete sequence of minerals shown in Bowen's Reaction Series is not seen in Skaergaard as amphiboles and biotite are absent. The intrusion exhibits differentiation to ferrogabbros rather than granites although there are minor quantities of acid rocks in the form of granophyre.

Figure 3.17 summarises the variations in chemical composition of the cumulus minerals. Plagioclase feldspar shows the most complete sequence varying continuously from An_{69} (i.e. 69% anorthite, 31% albite (see Fig. 3.3) at the base of the layered series to An_{33} at the top. Plagioclases of composition An_{77} have been found in the marginal border group, and it is assumed that this represents the composition of the first plagioclases to crystallise. If this is so, plagioclase feldspar of composition An_{77} should be found near

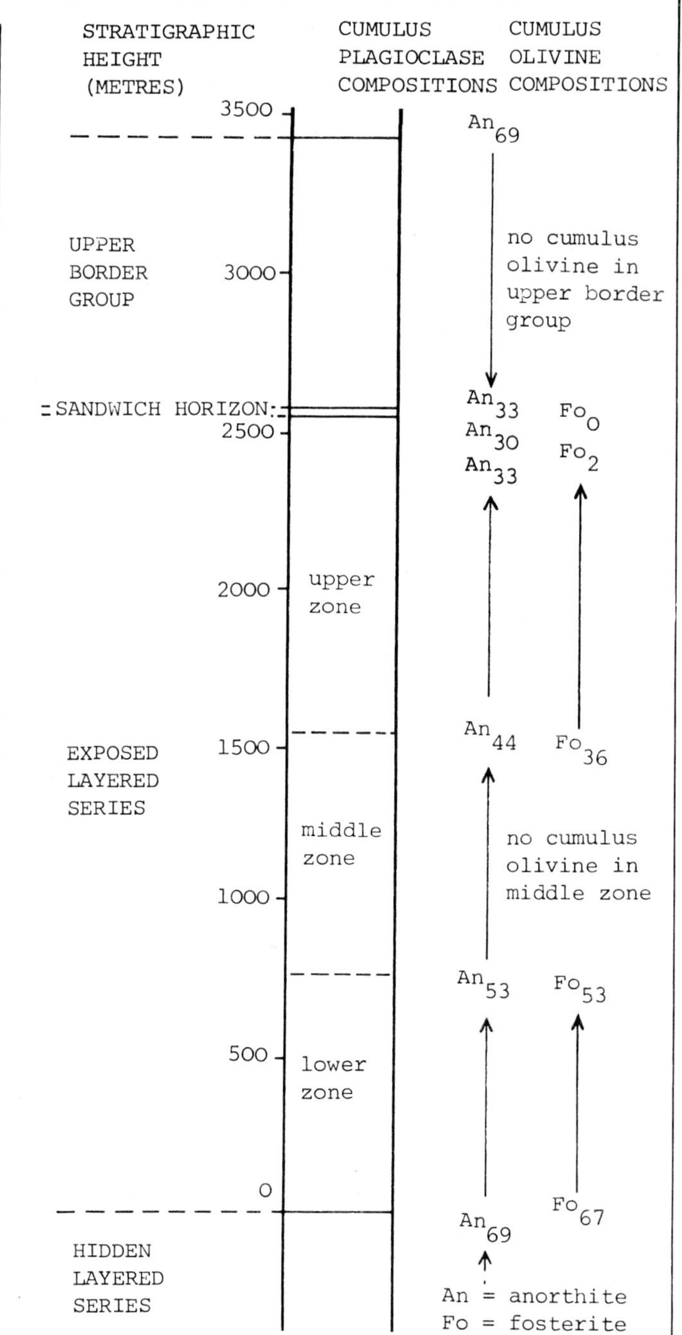

Fig. 3.17 Cumulus mineral composition within the Skaergaard Intrusion.

the base of the hidden part of the layered series. In fact, drilling has confirmed this prediction.

The sequence of chemical variations in the cumulate olivines is not complete since olivine is absent in the middle zone of the layered series. At the base of the lower zone of the exposed layered series the olivine has composition Fo_{67} (i.e. 67% fosterite, 33% fayalite), and it reaches Fo_{53} by the top of the lower zone. When it reappears in the upper zone its composition is Fo_{36} and by the top of the upper zone of the layered series it is Fo_2. When the top of the layered series is reached, the iron content of the rock is great enough for it to be called ferrogabbro.

The third part of the intrusion is the upper border group which, because of erosion, is only seen in part. This is thought to represent cooling from the top downwards since the reverse trend in plagioclase composition compared to that in the layered series has been recorded.

The last part of the intrusion to crystallise was the Sandwich Horizon between the upper border group and the Layered Series. This is the most acidic part of the intrusion and it contains lenses of granophyre with some quartz and orthoclase feldspar, set in quartz-ferrogabbros.

Thus, just as with the Palisades Sill, rocks of acidic composition can be obtained from a basic magma. A comparison of chemical analyses of Skaergaard (Table 3.3) with those given for the Palisades Sill (Table 3.2) shows similar trends although the final products of differentiation are more extreme in Skaergaard. However, in both cases, the amounts of acid rocks produced are very minor. Very slow cooling is necessary for this to happen and it has been calculated that over 12,000 years were needed for the exposed parts of the layered series alone to crystallise. Because of their slow cooling, each of these intrusions has preserved a record of the order of crystallisation of the minerals and has led to a greater understanding of the processes of differentiation working in large plutonic basic intrusions.

TABLE 3.3 CHEMICAL COMPOSITION OF ROCKS FROM SKAERGAARD INTRUSION (Weight %)

	1	2	3	4
SiO_2	47.92	46.37	48.27	58.81
TiO_2	1.40	0.79	2.20	1.26
Al_2O_3	18.87	16.82	8.58	12.02
Fe_2O_3	1.18	1.52	4.06	5.77
FeO	8.65	10.44	22.89	9.38
MnO	0.11	0.09	0.26	0.21
MgO	7.82	9.61	1.21	0.72
CaO	10.46	11.29	7.42	5.03
Na_2O	2.44	2.45	2.65	3.91
K_2O	0.19	0.20	0.34	2.39
H_2O+	0.41	0.29	1.13	0.21
H_2O-	0.10	0.09	0.37	0.19
P_2O_5	0.07	0.06	0.65	0.71
CO_2	0.06			
SrO	0.20			
BaO	0.02			
S	0.27			
	100.17	100.02	100.03	100.61

1 Olivine Gabbro from Chilled margin
2 Olivine Gabbro 500 m above lowest exposed horizon
3 Ferrogabbro 2,500 m above lowest exposed horizon
4 Granophyre from lensoid mass in upper part of intrusion

Granites

Granites and associated rocks such as granodiorite form the largest group of intrusive igneous rocks, comprising over 90% of the total. This contrasts sharply with extrusive rocks where basalt is the most common rock type. Part of the reason for this contrast between the composition of the main intrusive and extrusive rocks has been explained in a previous section on the crystallisation of magma, but part of the answer also lies in the origin of the respective magmas.

Basic magma is produced mainly by partial melting of the ultrabasic mantle. Does the acidic magma which forms granite also come from the mantle or does it have a separate source?

It is possible for small quantities of granitic rock to be formed by differentiation of the primary basic magma as has been illustrated in both the Palisades Sill and the Skaergaard Intrusion. However, in order to produce all the granitic rocks by differentiation, there would have to be at least ten times as much basic magma as acidic magma. This is clearly not the case, as the acidic intrusions are nearly twenty times as abundant as all other intrusions.

As basic magma rises through the continental crust, partial melting may occur and the products become assimilated into the magma, thus contaminating it. However, it is very unlikely that this process would result in any significant quantities of granites.

If most granites are not formed from primary basic magma, how do they originate? A clue to their origin is provided when the distribution of granites is studied. We find that the vast majority are associated with the continental crust, usually within former orogenic belts. The few small intrusions which are in oceanic areas can be accounted for by differentiation of a basic magma. The chemical composition of the upper continental crust closely approximates to that of granite, and the orogenic areas are regions where temperatures and pressures are abnormally high. Theoretically, therefore, granites could have been formed by the melting of the upper continental crust, but is this possible in practice?

Anhydrous (dry) granite begins to melt at about $950^{\circ}C$-$1,000^{\circ}C$ at the pressures encountered in the crust. As the geothermal gradient, at the time of orogenic activity, could have been as high as $30^{\circ}C\ km^{-1}$, the temperatures required to produce melting are found at a depth of less than 35 km, which is well within the continental crust in orogenic areas. Hydrous (wet) granitic melt can be produced at even shallower levels in the crust as the temperature required is only about 650°C, which is reached at a depth of 20-25 km. (See Fig. 3.6).

Although we can see that it is perfectly possible for granites to be produced by the heating up and eventual melting of continental crust, the idea has given rise to a long debate between geologists. On one hand there are the 'transformationists' who saw that granite could indeed be produced by transformation of continental crust. To the 'transformationists', granite is the end product of metamorphism, produced first as a 'mixed' rock, part gneiss, part granite, known as migmatite which itself undergoes complete transformation at a later stage, becoming granite. On the other hand there are the 'magmatists' who point to the many examples of granites that clearly have formed from an intrusion of magma, and are thus truly igneous. There are points to be made on both sides of the argument and the following examples illustrate features which suggest a migmatitic origin in some cases and an igneous, or magmatic, origin in others. There then remains the question of whether the genesis of these two types of granite can be connected in some way.

Migmatites and migmatitic granites

As has already been pointed out, it is very common to find granites in orogenic belts associated with areas of high grade regional metamorphism. The sequence of metamorphic rocks - slate → phyllite → schist → gneiss produced from shales or mudstones with increasing grade of metamorphism is well known. Not quite so well known, perhaps, are the properties of the group of rocks known as migmatites.

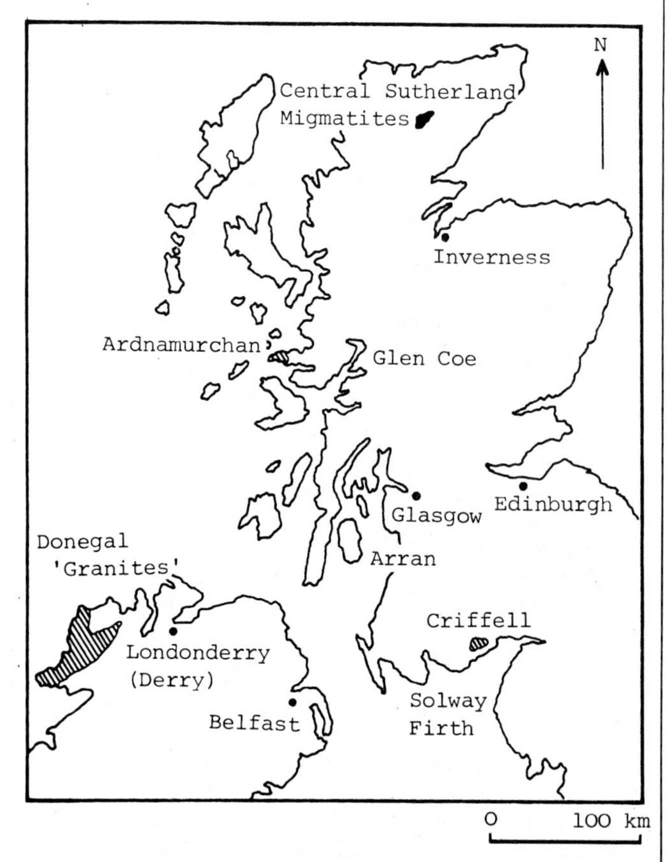

Fig. 3.18 Location of Central Sutherland migmatites and Donegal granites.

These migmatites are most often found associated with schists and gneisses and they are in fact 'mixed' rocks where granitic material has come to be mingled with a host rock which originated by high grade regional metamorphism. The granitic material seems to have migrated through the host, either as magma or aqueous fluids of granitic composition, although it is possible that it took the form of a diffusion of ions migrating through the pore fluids. These are rather loosely referred to as 'emanations'. It is often claimed that these emanations have their origin outside the host, perhaps rising from deeper levels in the crust, although other authors believe that in some cases the granitic material may have originated as a segregation from the host itself.

It is quite easy to see that there are two distinct components in hand specimens, even though metasomatic reactions between them will have caused some modifications in composition. The original composition of the host will have an important part to play in the composition of the final product, but with increasing migmatisation a rock is produced which approaches granite in composition.

Therefore, it has been quite convincingly argued that migmatites provide a link between the true metamorphic rocks and granites and that the complete metamorphic series should read: slate → phyllite → schist → gneiss → migmatite → granite. In Britain the main areas where migmatites are found are in the Highlands of Scotland and in Ireland (Fig. 3.18). Here, during the Caledonian Orogeny, a broad zone of regional metamorphism was formed and migmatites can be seen in such places as the Central Sutherland Complex in Scotland. Here, as in many other examples there is a gradation from an area where the host rock, in this case Moine Schists, is traversed by a few veins of granite, through zones where the granitic component becomes more conspicuous to local occurrences of almost pure granite conforming to the strike of the country rock (Fig. 3.19).

The Main Donegal granite of Ireland shows evidence that it was emplaced by a lateral wedging process yet it has included in it rafts of the original country rock which are not only aligned parallel to the regional trend but also preserve the original or 'ghost' stratigraphy of the area across the granite outcrop (Fig. 3.20). Although, therefore, this is not a migmatitic granite in the sense that it has formed 'in situ', clearly the magma has not moved far from where it originated and it does not show the homogeneous nature which might be expected of a thoroughly magmatic granite.

Similarly, the older granodiorite, north west of the Main Donegal granite, contains inclusions of Dalradian migmatites which have retained, to a great extent, their original position. The granodiorite appears to penetrate the country rock but sufficient displacement of the xenolithic blocks has been noted for a magmatic origin to be postulated for the granodiorite.

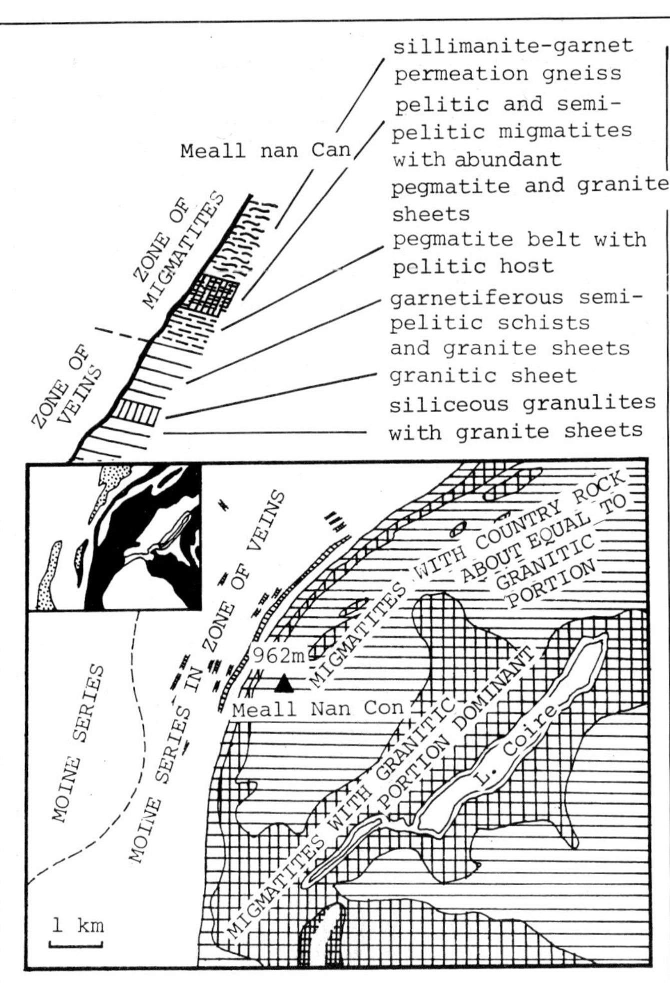

Fig. 3.19 Regional migmatisation, Central Sutherland. The main map shows the zones of migmatisation; the inset map shows the distribution of original lithological types, black pelites, dotted hornblende-gneiss, unornamented psammites. The section of Meall nan Con (Ben Klibreck) above shows zones of migmatites over zones of veins. The height of the section is about 800 metres and the vertical and horizontal scales are the same.

In contrast, the remaining two granite outcrops in the area are very clearly magmatic, although of different form. The Rosses granites form a ring complex and the Ardara granite is a diapiric intrusion. These different types of intrusions are described later.

An example of a magmatic granite

Many examples of granites which are clearly of magmatic origin could have been chosen from the British Isles; indeed two have already been mentioned in the previous section. In general, they tend to be smaller, more homogeneous bodies than the granites that are believed to have been generated 'in situ' by the processes of migmatisation.

Although not strictly a granite, because of its lower silica content, the main part of the Criffell granodiorite of the Southern Uplands of Scotland provides an excellent example of an intrusion. It is one of a series of Caledonian granite outcrops in the area and it has been forcefully intruded into folded Silurian shales and greywackes. There are three parts to this intrusion, for the main granodiorite grades into a porphyritic

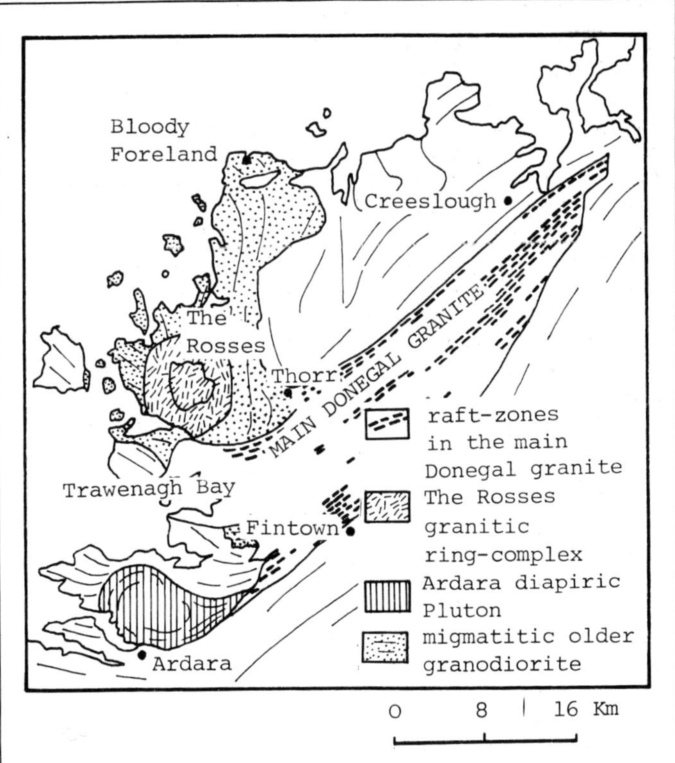

Fig. 3.20 Diagrammatic map of the Donegal Granite.

granodiorite with feldspar phenocrysts in its central portion. The third part is older than the main complex and is seen at its south western end. There, finer grained granodiorites are associated with some quartz-diorite rocks (Fig. 3.21).

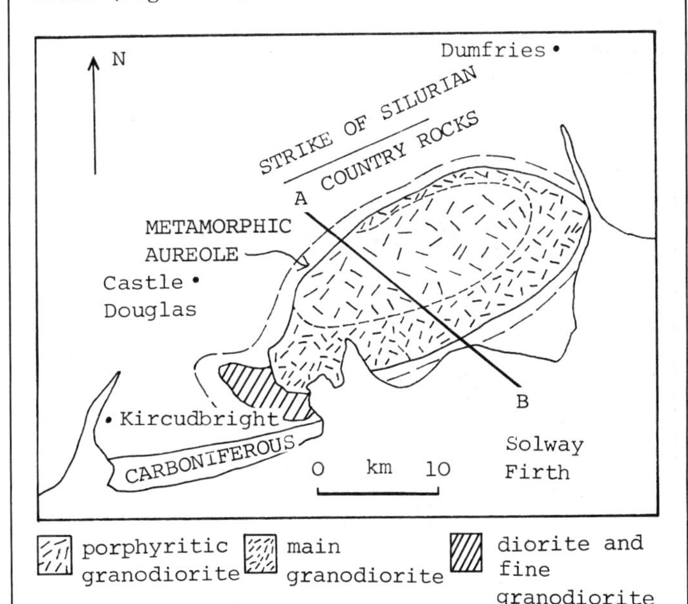

Fig. 3.21 The Criffel granodiorite complex.

The Criffell granodiorite shows a large number of features which could only have originated by the intrusion of hot magma into colder country rocks. Most significantly, in contrast to the migmatitic granites previously described the contacts with the country rocks are sharp and they dip steeply outwards. This has been confirmed by geophysical surveys across the intrusion. Figure 3.22 illustrates the interpretation by one of the authors after extensive geophysical work in the area. (See the Unit *Geophysics* for further details of the technique.)

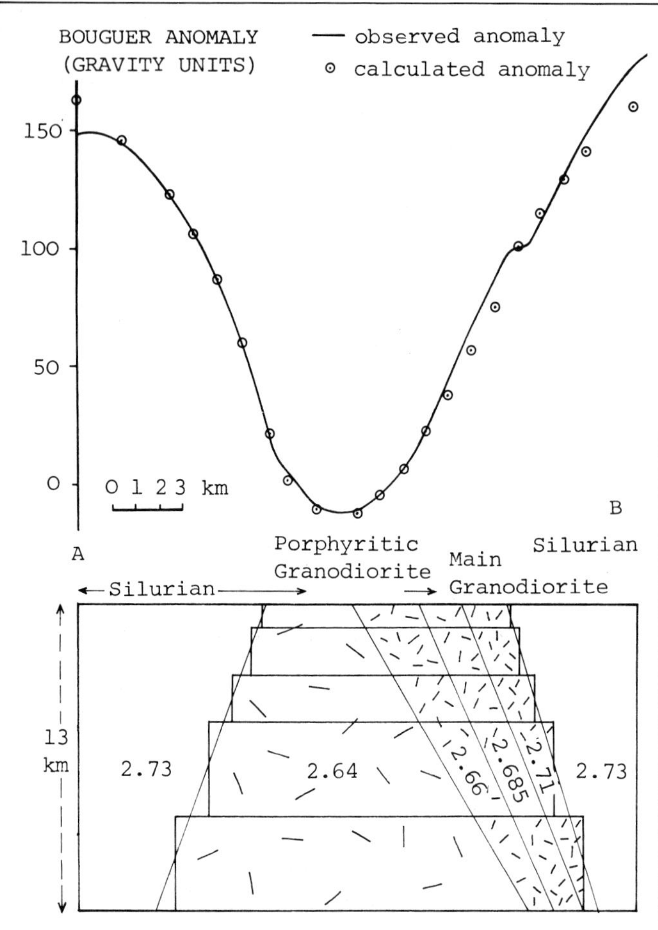

Fig. 3.22 Section across a three dimensional model of the Criffel granodiorite showing a good fit between the observed and calculated gravity anomalies. 1 gravity unit = 10^{-6} m s^{-2}. 2.73 etc = rock density in tonnes m^{-3}

There is also no gradual transition from high grade regional metamorphic rocks or migmatites, since the country rocks have only previously been metamorphosed to slate grade. The intrusion has, however, formed a clear metamorphic aureole within the country rocks which have been hornfelsed and in some places mineralised. In the outer parts of the intrusion xenoliths are common and veins of the granodiorite invade the country rock in places. A flow foliation has been recorded in the outer parts of the intrusion which indicates that the origin must have been from a mobile magma. Although the outcrop of the granite is slightly elongated in the direction of the strike of the country rocks it is completely discordant to it at its western end which again contrasts with a migmatitic granite.

All these features make it clear that this particular intrusion originated from a hot, mobile magma. It is thought that the mass intruded to a relatively high level in the earth's crust, perhaps rising to within 1,000 m of the earth's surface before finally solidifying.

The granite series

The examples illustrate that some granites result from crustal melting - the so-called 'metamorphic' or migmatitic granites - whilst others have a definite magmatic source - the so-called 'igneous' granites. How can these two greatly contrasting origins be resolved, if indeed they can be resolved at all?

In his book *The Granite Controversy* (1957) H H Read summarised the arguments and proposed what he called the 'Granite Series' which provided a link between the two types of granite. In this series, the metamorphic granites represent deeper levels in the earth's crust where partial melting has taken place, a process known as anatexis. The melt has then consolidated 'in situ'. Sometimes, this melt becomes mobile and, because it has a relatively low density compared to the rocks around it, it will then move upwards to intrude into the higher levels of the crust where it will crystallise to form a magmatic granite. Figure 3.23 summarises the various stages and links between them in the Granite Series.

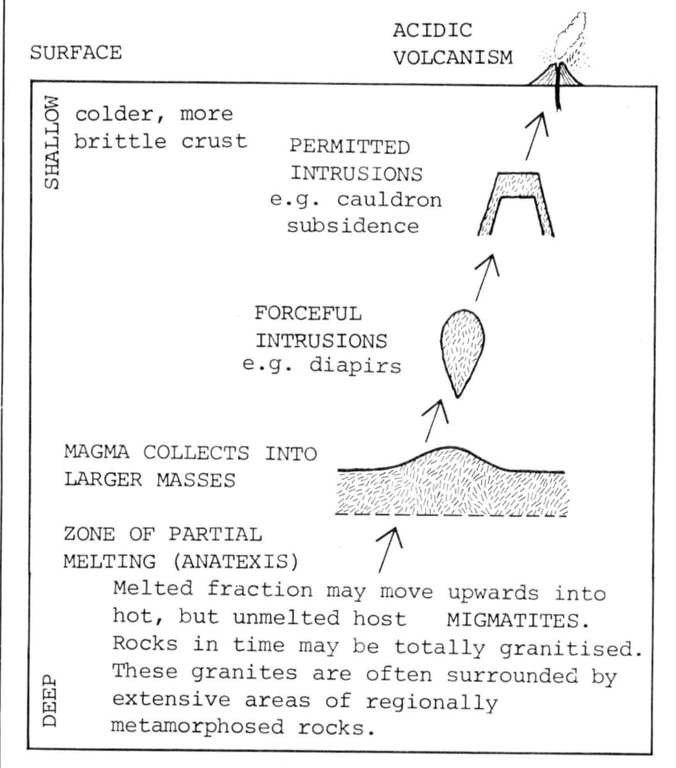

Fig. 3.23 The Granite Series.

One question needs consideration. If most granites can be referred back to crustal melting, how is it that they show little variation in composition when the crust is very variable? Experiments on mixtures of quartz, orthoclase and plagioclase have shown that they begin to melt at varying temperatures, depending upon the relative proportions of the minerals in the mixture. The mixture which remains liquid to the lowest temperatures contains approximately equal proportions of the three components. Analyses of a great number of natural granites, which were poor in ferromagnesian minerals, showed that they contained quartz, orthoclase and plagioclase in much the same proportions as in the artificial mixture. This is because the natural granites will contain the most easily melted fraction, i.e. melted at the lowest temperatures, of the original crustal rocks and so variations in the composition of the crustal rocks will be evened out. Figure 3.24 shows how well the low temperature mixtures match the maximum for the granites.

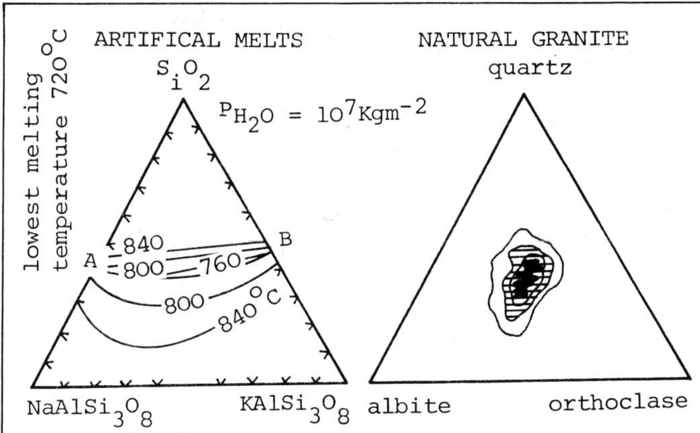

Fig. 3.24 The system SiO_2, $NaAlSi_3O_8$, $KAlSi_3O_8$ and its application to natural granites. (a) The temperatures at which crystallisation begins in a quartz-orthoclase-albite liquid, in the presence of water, at a water-vapour pressure of $10^7 Kgm^{-2}$. Above AB, quartz appears first, below AB feldspar first. The lowest temperatures of crystallisation are shown by mixtures having a composition in the central part of AB. (b) Composition of 571 natural granites containing less than 20% of ferromagnesian minerals, expressed in terms of quartz, orthoclase and albite. Note that the maximum falls in the position of the low-temperature mixtures.

Granites are also found outside orogenic areas where the thinner crust and the lower geothermal gradient means that it is not possible for them to have been produced by crustal melting. For these granites we have to revert to postulating an origin in the mantle. The peridotite mantle must have suffered extreme differentiation to produce a granitic melt. This is theoretically possible, although very large quantities of material would be required, the ratio of original peridotite to granitic melt produced being approximately 100 to 1. Evidence for this hypothesis has recently been found. In crustal granites, and those thought to originate from the mantle, the ratios of two isotopes of strontium have been found to be different, reflecting their different origins. This new work on the strontium ratios has led to increased speculation about the origin of granites. The migmatitic granites are obviously formed in the crust as they are derived from partial melting of crustal rocks. However, magmatic granites could either be derived from partially melted crustal material which becomes mobile and moves upwards as outlined in the Granite Series above or they could alternatively have been derived from the mantle.

The space problem

Several very large granitic masses are known from different continental areas. These are called batholiths and may be hundreds of kilometres in length (Fig. 3.25). Smaller offshoots from the main batholith can occur and these are known as stocks or bosses.

When granites can be inferred to have been formed 'in situ' by an intensification of metamorphic processes leading to crustal melting, there is no difficulty in accounting

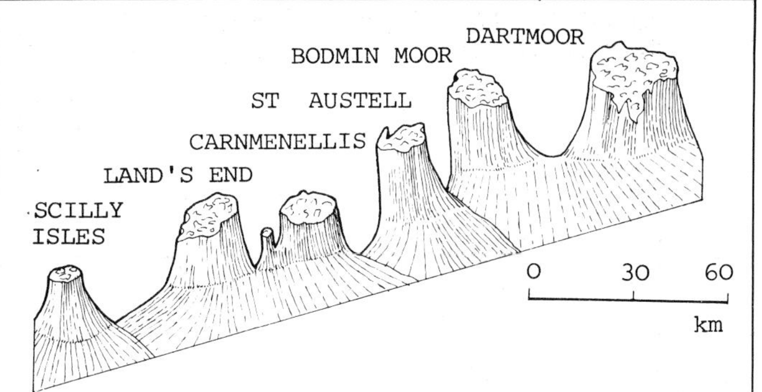

Fig. 3.25 The granite batholith of south west England. The main batholith has a number of smaller offshoots known as stocks or bosses which outcrop at the surface.

for their vast size. The original rocks have merely been altered or transformed. However, when it can be demonstrated that large masses have been formed by the intrusion of magma which came from somewhere else, a mechanism or mechanisms have to be found which can account for the displacement of the existing country rocks. These mechanisms fall into two broad groups known as <u>forceful</u> and <u>permitted</u> intrusions. Forceful intrusions make room for themselves by shouldering aside the country rocks whilst permitted intrusions are formed more passively by magma rising to occupy spaces left by detached masses of country rock. The same principles apply to large intrusions of any composition.

Forceful intrusions

<u>Diapiric intrusions</u> The density of the hot magma tends to be less than the surrounding rocks and so this magma can flow slowly upwards pushing the existing rocks out of the way. In the most simple of cases the rocks are merely bent around the granite forming a <u>dome</u> (Fig. 3.26a) but further upwards motion of the magma will rupture the country rocks and form a <u>diapir</u> (Fig. 3.26b) such as in the Ardara granite of Donegal.

The mechanism of intrusion can usually be inferred from associated structures in the country rocks such as folds formed by the upward and outward pressure of the magma or simply by the dip of the country rock which tends to incline outwards from the granite margin (Fig. 3.27).

These types of intrusion will tend to form 'blisters' on the surface which will make them the target for erosion and so the overlying cover of country rocks can be quickly removed.

Other types of intrusions which have forcefully pushed aside the country rocks include the large, concordant, sheetlike bodies of laccoliths and lopoliths. Laccoliths cause an arching of the country rocks, much like domes, but less strongly arched. They are more common with the acidic rocks than the basic rocks since the more viscous nature of acidic magmas restricts their flow along the bedding planes and so a thickening occurs around the feeder dyke (Fig. 3.26c).

Lopoliths are mainly associated with basic magmas and are large, saucer-shaped

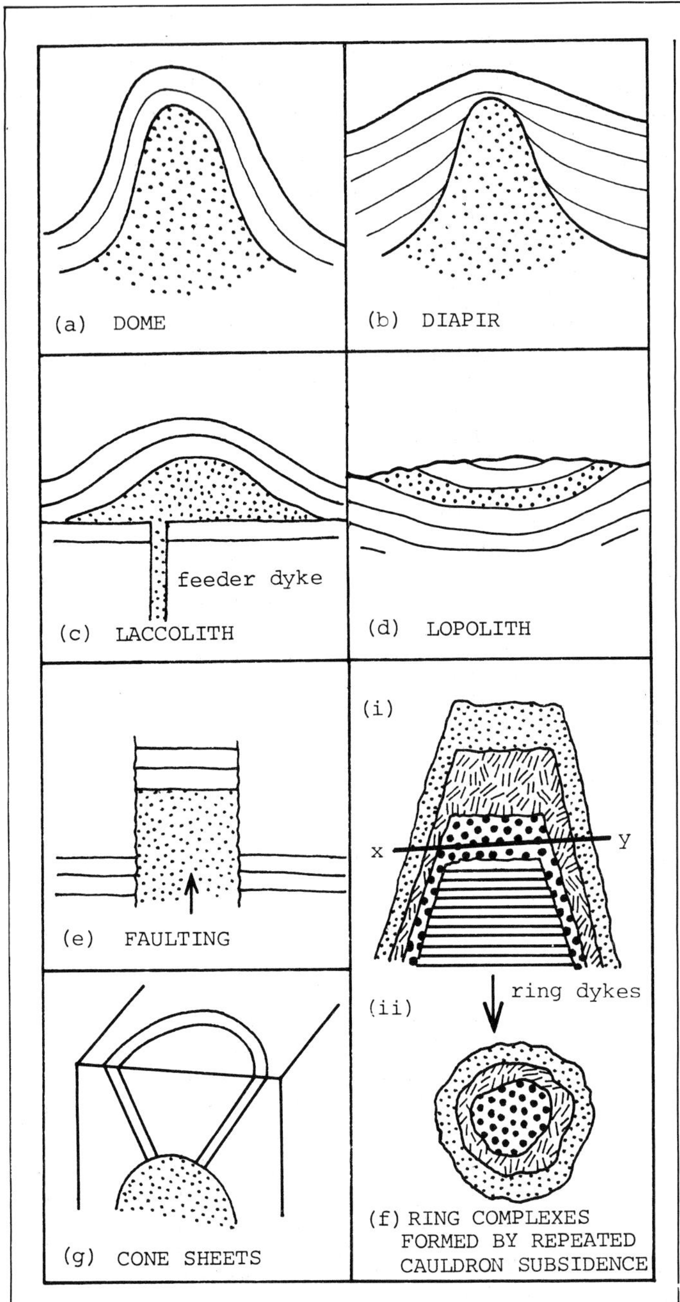

Fig. 3.26 Simplified diagrams to illustrate the various types of intrusions.

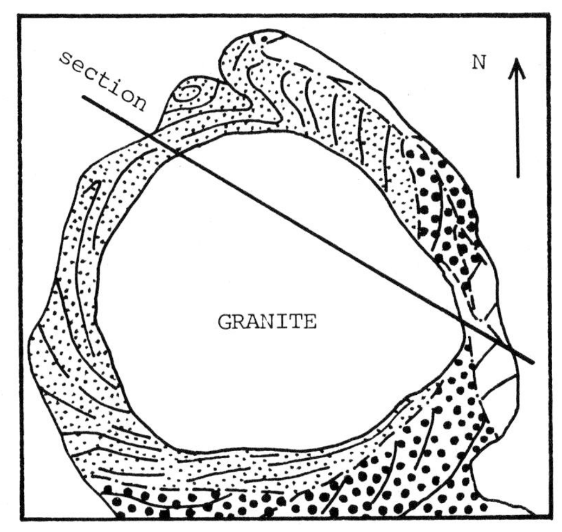

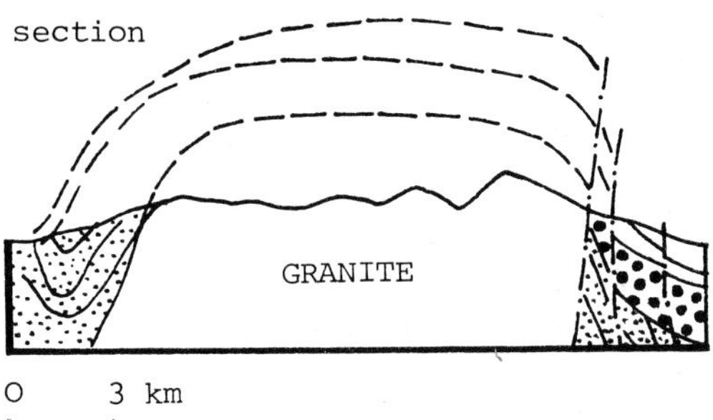

Fig. 3.27 Structures produced in the envelope rocks by the intrusion of the Arran granite dome.

intrusions. Most appear to have had complex cooling histories and are not, as was first supposed, merely a result of the crust sagging under the excess volume of magma. A very large, well-known example of a lopolith is the Bushveld Complex of South Africa (Fig. 3.26d).

Faulting. Where there are parallel faults, the intrusion makes room for itself by pushing up the country rocks between the faults (Fig. 3.26e).

Permitted intrusions

Cauldron subsidence There is another more elaborate method by which magmas can be emplaced by faulting. A ring fracture forms, but instead of the central block being pushed up it subsides into the magma and the magma squeezes up the sides of the block, acting as a lubricant helping the block to sink. The magma then occupies the cauldron so formed (Fig. 3.26f). These types of permitted intrusions are relatively common and good examples can be seen in Scotland. In fact the process can be multiple with the cauldron sinking repeatedly. At Glen Coe the older Moor of Rannoch granite has been intruded by a second, the Ben Cruachan granite. In the ring complexes of Ardnamurchan in Scotland and the Rosses Complex of Ireland there have been three phases of emplacement by this mechanism (Fig. 3.28).

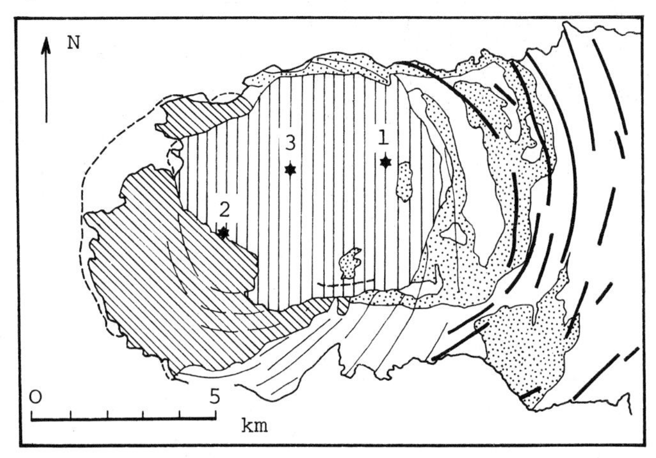

Fig. 3.28 Ring complexes of Ardnamurchan. The positions of igneous centres have migrated.

Stoping happens on a smaller scale and enables the magma to digest its way upwards. Lumps of wall and roof rock fall into the magma and they become homogenised with it permitting the magma to move upwards into the space that is left. Evidence of this process can remain as the magma may have a large number of xenoliths in it, where blocks of country rock have not been completely assimilated into the magma.

Minor intrusions

The two most usual types of minor intrusions are sills and dykes. Sills are concordant intrusions, that is, they are sheet-like structures which follow the bedding or other planes of structure. A sill may cross from one plane to another, in which case it is said to be transgressive. There is great variation in the size of sills, the Palisades Sill at over 300 metres thick being a particularly large example. One of the best-known British examples is the Great Whin Sill of Northern England. This averages approximately 30 metres in thickness and it underlies a vast area from the Farne Islands in the north to the Tees Valley in the south. Perhaps it is most well known where its outcrop has formed a north-facing scarp slope along which Hadrian's Wall has been built. Many sills do not reach this size and are found as thin intrusions of no great lateral extent (Fig. 3.29).

Fig. 3.29 Megaligger Rocks, South Cornwall. Acid sills in shales.

Dykes are discordant structures where a sheet of magma has cut vertically or nearly vertically across the bedding planes. Dykes can vary greatly in thickness, from a few centimetres, to hundreds of metres across and they can either be of restricted length or run for considerable distances. For instance, the Cleveland Dyke can be traced intermittently all the way from Arran, across northern England to the North Yorkshire Moors. In the Tertiary Igneous Province of north-west Britain, numerous dykes can be seen running parallel to one another in dyke swarms.

A number of minor intrusions are often associated with the major intrusions described above. Commonly, the stress of the magma intrusion will cause fractures in the country rocks and the minor intrusions fill these cracks. A radial pattern of dyke swarms may occur or there may be a concentric pattern known as cone sheets. The cone sheets dip inwards towards the major intrusions with which they are associated (Fig. 3.26g). Figure 2.2. is a photograph showing a cone sheet in the Cuillin Hills on Skye. A radial dyke swarm has originated on Tristan da Cunha (Fig. 1.4) due to stresses built up as magma collected beneath the volcano.

One problem often confronting geologists when they see a layer of igneous rock lying conformably in a succession of sedimentary strata, is to tell whether it is a sill or a lava flow. Perhaps the easiest way to decide is to observe the baked area formed. A sill will have a baked area above and below it but with a lava flow one can only develop in the underlying rocks as there was nothing above it at the time of the extrusion.

Further indications are given by the nature of the upper surface. With a lava flow, this may be uneven and have a soil developed upon it as it could have been exposed to subaerial processes of weathering. On the other hand, a sill could not show this, as it is by definition intrusive. It could, however, contain xenolithic inclusions of the overlying strata whereas it would be impossible for a lava flow to have these.

4
The Global Scene

The Origin of Basic Magma

As we have seen in the previous section, granitic rocks are usually formed from material which is generated within the continental crust, but this is not the case with basic magma. Basalts are frequently found erupting in the ocean basins, far removed from continental crust, and basic magma also tends to be much hotter than acid magma. It usually erupts at temperatures of over 1,000°C whilst it is rare for an acidic lava to exceed 800°C. These higher temperatures are very infrequently reached within the crust. Basic magmas are therefore considered to be derived from the mantle.

For obvious reasons, our knowledge of the mantle is somewhat limited. However, it is possible to obtain information about it in a number of ways. First, xenoliths of mantle material are brought up in kimberlite pipes. These pipes are roughly cylindrical tubes which cut through the continental crust in various parts of the world, notably in South Africa. They are formed of ultrabasic rocks and, as well as the mantle xenoliths, some contain diamonds which show that the pipes must have originated at least 200 km down in the mantle where there is sufficient pressure for diamonds to form.

Further information about the mantle is gathered from the limited number of localities where rocks from it are actually exposed. The closing of the African plate relative to the European has caused the crust in between to be squeezed and buckled. This has become so intense in places that a small part of the upper portion of the mantle has actually been thrust up, folded and exposed at the surface in the Troodos Mountains of Cyprus.

Other evidence concerning the properties and composition of the mantle is obtained indirectly by a variety of geophysical methods such as its response to seismic waves. This reveals a lot about its density and its elastic properties. From the various lines of evidence, it is clear that the upper mantle is a solid with an ultrabasic composition. Originally, it was thought that the upper mantle had a fairly homogeneous composition, but as our knowledge has grown it is seen to be increasingly complex. Nevertheless, fundamentally it is composed of peridotite.

It is from this peridotite that the primary basic magma is thought to originate by partial melting where the mantle temperature becomes locally raised. Even though the upper mantle's composition is variable, the primary basic magma seems to have a reasonably uniform composition. This is related to the way in which partial melting takes place. Once partial melting has started, it proceeds without delay and is able to mix after only a little melting has taken place which helps to produce a homogeneous magma. This melt has a surprisingly low viscosity. Under a pressure of 25×10^8Pa, which is that encountered at the depths where the magma originates, the viscosity is only 25 poises. For comparison, glycerine at atmospheric pressure has a viscosity of 100 poises. This melt is therefore very mobile.

The depth at which this partial melting takes place is variable, but a zone which lies at a depth of around 70-150 km, known as the asthenosphere, seems one likely source. This zone has been recognised from seismic studies because the velocity of shear waves reaches a minimum as they pass through it. This suggests that it is a weak zone in which, because of abnormally high temperatures, rocks are close to their melting points. It is even possible that it contains small amounts of liquid. Above the asthenosphere, the upper mantle and crust are more rigid and are known as the lithosphere. The mechanically weak asthenosphere is probably the area in which the lithospheric plates detach from the lower mantle.

This primary basic magma can become modified on its journey to the surface, leading to a variety of different volcanic rocks. There is a broad regional pattern to this variation which will be considered in the next section.

Petrographic Provinces

When the pattern of modern igneous activity is examined, it is very easy to see that there is a regular distribution, the volcanoes usually being associated with active plate margins (Fig. 4.1).

Our earlier studies of Mount St Helens and Tristan da Cunha showed that the type of igneous activity and the nature of its products varies in different parts of the world. However, the pattern of activity is certainly not random and a number of petrographic provinces (also known as magmatic or volcanic provinces) have become recognised. A petrographic province is a broad region where volcanic rocks occur which are genetically related and which belong to a similar period of activity. Each of these provinces

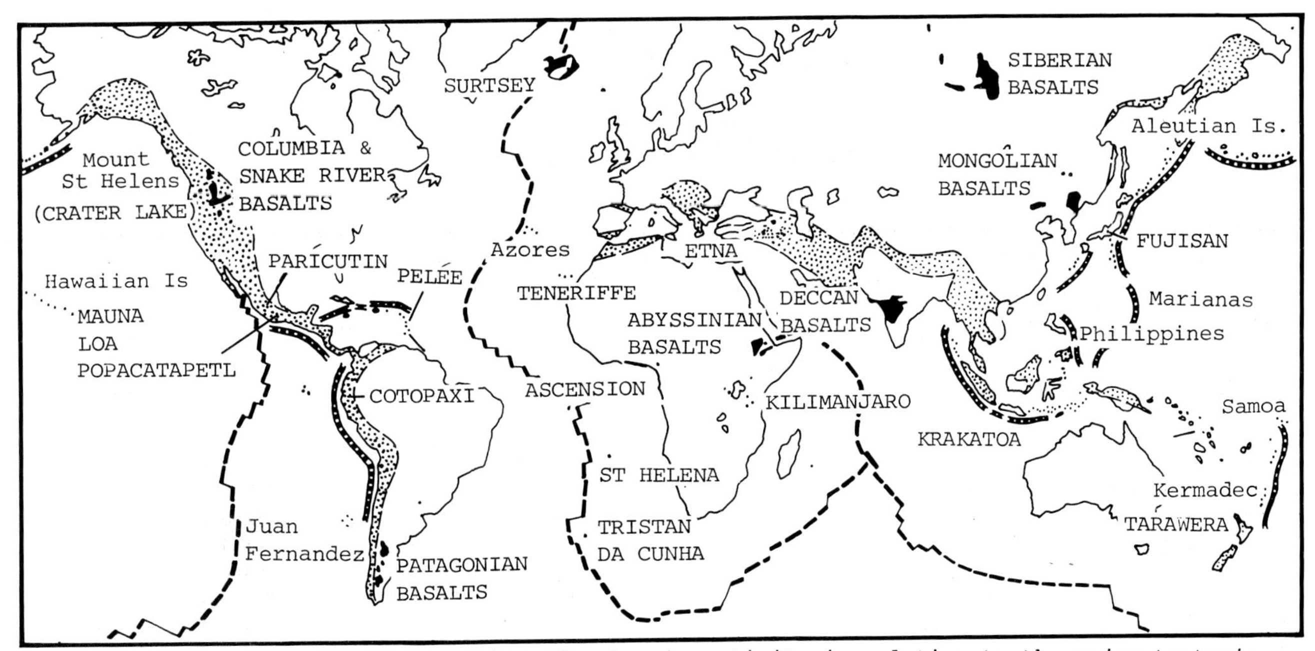

Fig. 4.1 Map showing the distribution of volcanic activity in relation to the major tectonic features of the earth. Positions of the crests of oceanic ridges are indicated by a heavy line, broken to show displacements due to transverse faults. The relationship between oceanic trenches (black and white ornament) and the belts of Tertiary to Recent folding of the continents and island arcs (stippled) are indicated. Dots indicate the positions of only a selection of Recent or currently active volcanoes. The main areas of Tertiary or young flood basalts are shown in black.

will have a limited 'life'. For example the distribution of igneous activity in the Mesozoic would show a very different pattern from today. Figure 4.2 is a schematic diagram showing how magma is generated today within the general framework of plate tectonics, the differences in generation leading to the modern petrographic provinces.

Active ocean ridges

The first province of modern igneous activity we shall consider is the ocean ridge system. This world-wide system of ridges is the site of numerous volcanoes, both submarine and those which reach above sea level and form volcanic islands such as Tristan da Cunha. The ocean ridges are the constructive plate margins. Here, rising convection currents in the mantle make them areas of high heat flow. Partial melting is therefore accomplished at relatively shallow depths. The mid ocean ridge basalts are tholeiitic where the higher silica content is a reflection of partial melting at shallow depths. As the majority of these lavas are erupted under water, they typically show a pillow structure.

North Atlantic Tertiary Province

In the North Atlantic, the ridge system is still very active and visible above sea-level in places like Iceland. However, in north-west Britain and Greenland there is substantial evidence that these areas were also active when the Atlantic was first opening up in the Tertiary and when, as a consequence, they both lay close to the ridge system. Because of this link between Greenland, Iceland, the Faeroe Islands and North West Britain, they are usually collectively known as the North Atlantic (Brito-Arctic or Thulean) Tertiary Igneous Province.

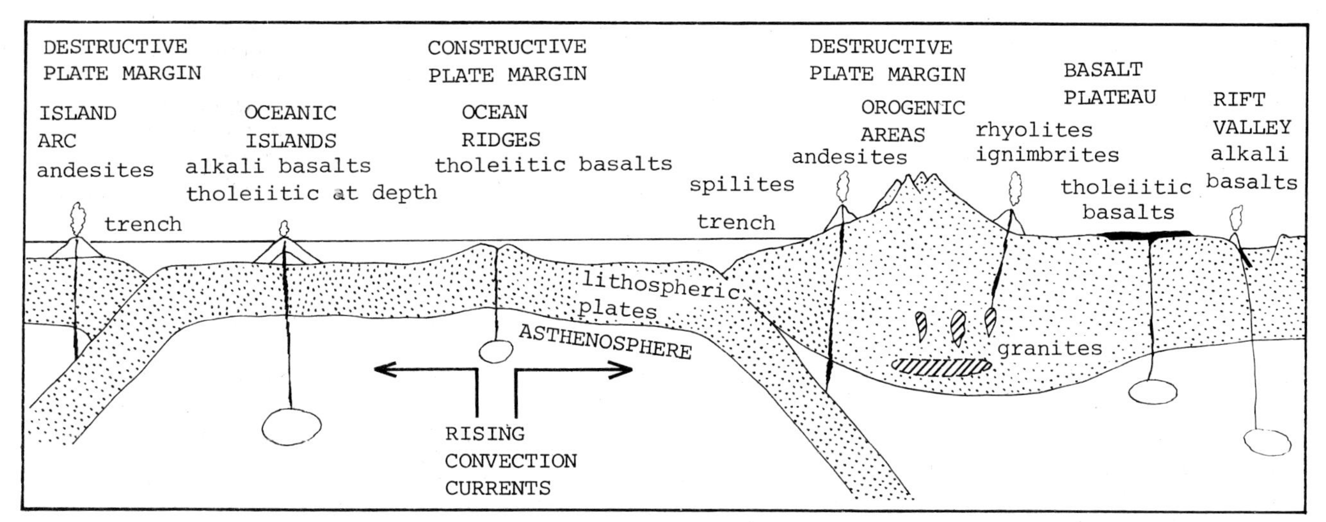

Fig. 4.2. Magma generation and plate tectonics.

Oceanic islands

There are other volcanic islands in the ocean basins which are not connected with the ridge systems, notably the Hawaiian Islands. Here, activity seems to be associated with a 'hot-spot' in the mantle. When the ages of the rocks forming these islands are examined, it can be seen that the dates of activity decrease from the north-west to the south-east, which could be related to the movement of the Pacific Plate to the north-west over such a 'hot spot' in the mantle (Fig. 4.3).

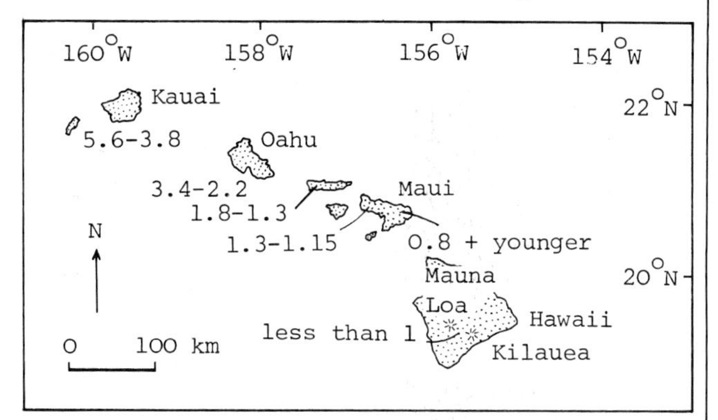

Fig. 4.3. Potassium-argon ages of the Hawaiian Islands in millions of years.

Recent activity, forming lavas in the upper part of the islands, has produced alkali-basalts whilst the vast bulk of the shield volcanoes is tholeiitic. The reasons for the differences in the lava types stem from changes in the magma reservoir. The original tholeiitic lavas may have been produced in areas of higher heat flow, like those found close to the ridge crests but, with time and increasing distance from the ridges, areas of lower heat flow are entered and the lavas would have to be erupted from greater depths resulting in the alkali basalts.

Although the actual mechanisms of generation are still unclear, much detailed seismic work has been completed on Hawaii which gives us information about the rise of magma to the surface. These studies have revealed magma collecting at a depth of 60 km below the surface and over a period of months its slow rise to the surface has been followed. It collected at 2-3 km below the surface just before eruption, causing an expansion of the summit, but after eruption rapid deflation occurred.

Spilites

One other group of igneous rocks associated with the oceanic areas needs to be considered. These are the distinctive basalts known as spilites. Near the continental margins where the basaltic pillow lavas of the oceanic crust are associated with thick accumulations of sediment, the original composition of the basalt has become altered. They seem to have been affected by the introduction of sodium, possibly derived from the sea water trapped within the sediments. The main feature of the spilites is that the basic calcium-rich plagioclase feldspar has been replaced by the sodium-rich variety of plagioclase feldspar, albite. Chlorite and actinolite also replace some of the original pyroxenes.

Spilites are often found exposed closely associated with ultrabasic rocks. This assemblage is known as an Ophiolite suite. The ultrabasic rocks may represent the top part of the mantle overlain by basaltic oceanic crust, the spilites. This means that when ophiolites are found as in the Troodos Mountains of Cyprus, we may be getting an insight into mantle structure.

Destructive plate margins

At destructive plate margins another distinctive suite of volcanic rocks arises. Typically, andesitic lava is erupted at these margins as a plate is forced down along the subduction zone into the mantle, with its line of descent marked by a trail of earthquakes known as the Benioff Zone. Gradually, the plate will be warmed up as it is forced deeper into the mantle and enormous amounts of heat will also be generated by friction. This will cause partial melting of the top section of the lithospheric plate, the basaltic oceanic crust. As we discovered in a previous section of this Unit, a consideration of Bowen's Series shows that the silicate minerals melt in a definite order. Partial melting can therefore lead to a different product from that formed by complete melting. In this case, andesite is produced by the partial melting of basalt.

The destructive plate margins which surround the Pacific Ocean give rise to a distinctive area of andesitic vulcanicity, the so-called Circum-Pacific Belt or Pacific Ring of Fire. In fact, a boundary between the dominantly basaltic lavas of the ocean basins and the andesitic vulcanicity surrounding the ocean known as the Andesite Line is recognised. The differences between the two petrographic provinces are very significant as they reflect contrasting genesis of the magma. Also, the andesitic vulcanicity tends to be more explosive than that associated with the basaltic lavas.

Andesites erupt on the surface in two situations. Where the plate is subducted beneath oceanic crust, a chain of volcanic islands, known as an island arc, is formed. These curved chains of islands are seen in such places as the Caribbean and Aleutian Islands. Andesitic volcanoes also erupt on the continents and the classic shaped cones typical of this type of vulcanicity are seen in the Rockies and in the Andes where again a plate is being subducted, but this time beneath a continental rather than an oceanic plate. Mount St Helens, referred to in the introductory section, is a volcano of this andesitic type. As a large number of andesites are found in association with continental crust, it was thought the andesites may have originated by contamination of the primary basic magma as it rose through the more acidic crust. Whilst this idea has been largely superseded by plate tectonic theory, it is probable that the contamination process is still responsible for some of the andesites and rhyolites encountered in orogenic areas.

Although andesites are the dominant lava erupted in the petrographic province, basalts are also seen where partial melting has proceeded to a greater extent. Granite batholiths and their associated volcanic products such as rhyolite and ignimbrite are also associated with orogenic areas. However, their probable origins by the melting of contenental crust has already been discussed in some detail in a previous section of this Unit.

Rift valley vulcanicity

By no means all volcanic activity is confined to the ocean basins and margins and there are many extensive areas of volcanic activity in continental regions. The rift valley system of East Africa is a modern petrographic province in which the lavas are distinctively alkaline. The lavas include trachytes and alkaline basalts which contain a greater proportion of silica-deficient minerals, such as nepheline, than normal basalts.

The rift valley may be the site of a future continental split or, as seems more likely, it is an attempt at splitting that failed. The distinctive lavas of this province originate in the upper mantle and make their way to the surface along the fault zones of the rift valley.

Plateau basalts

Also in the continental areas, although not modern petrographic provinces since they have no active counterpart today, are very extensive areas of plateau basalts. Here, vast piles of tholeiitic basalt have built up, erupting from a series of fissures. They are seen in the Deccan Plateau of India where over half a million km^3 of basalt is found and in the Columbia-Snake plateau of the north-western United States. In Britain, the Antrim plateau is also of this type.

The composition of these basalts is very similar to that erupted at the ocean ridges and so similar conditions of high heat flow and relatively low pressure probably existed. That these basalts erupted from fissures shows that they occurred in areas of crustal tension, as indeed exists at the ocean ridge crests. It has been suggested that these outpourings were started by the initial activities of the spreading ridges. The magma generated spilt out onto the continents before the oceans opened up. The position of many of the lava plateaux on the margins of the present continents tends to support this theory (Fig. 4.4).

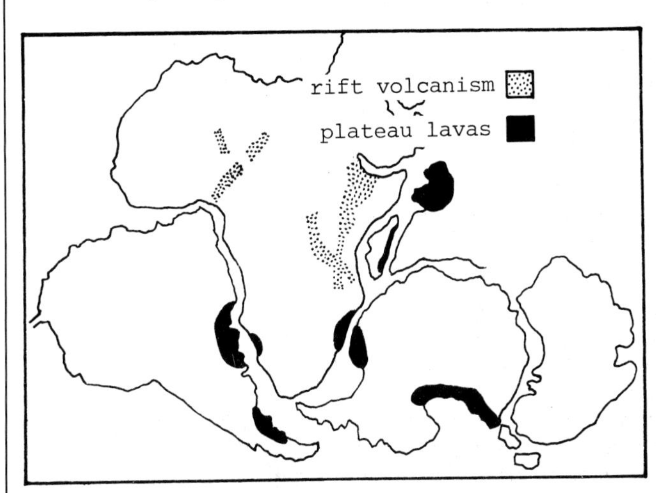

Fig. 4.4 Reconstruction of southern continents showing distribution of regions affected by Mesozoic and younger tholeiitic plateau basalt volcanism and some rift valley features characterised by alkali basalt volcanism.

5

Past Igneous Activity in the British Isles

Although there is no igneous activity in Britain today, there was a great deal in the past. As was dicussed in the previous section, modern igneous activity is largely confined to active plate margins and it therefore seems logical to assume that evidence of igneous activity in Britain in the past must usually be related to times when it was close to a plate margin.

Table 5.1 shows when the main periods of igneous activity are found in the British stratigraphic record. The nature of the activity can then be related to the modern petrographic provinces so that the type of plate margin causing the activity can be deduced.

It is very difficult to reconstruct what plate motions may have been like in the Precambrian and it is likely that, in the early stages of the formation of the crust, the plate tectonic activity was not as we know it today. Since then, however, a fairly clear pattern of Britain's plate tectonic history emerges.

Throughout the Lower Palaeozoic there was closing of the Iapetus Ocean, with northern Britain being on the same side as Greenland and North America, and the rest of Britain on the other side. Following the closing of this Ocean, which is marked by the Caledonian orogeny, Britain became part of a stable plate, continental to begin with but later covered by shelf seas. In the Upper Palaeozoic, the main phase of igneous activity was connected with the Midland Rift Valley of Scotland, although the southern part of England was affected as the ocean to the south of Britain closed causing the Armorican (or Hercynian) orogeny. In the Mesozoic, igneous activity was more restricted as Britain remained part of a stable plate. The activity was connected with the break-up of the northern continents. The first attempt at splitting seems to have been in the North Sea but eventually the con-

TABLE 5.1 AN OUTLINE OF PAST IGNEOUS ACTIVITY IN THE BRITISH ISLES.

PERIOD	LOCALITY	NATURE OF ACTIVITY	PLATE TECTONIC INTERPRETATION
QUARTERNARY AND TERTIARY	W. Scotland and N. Ireland	basalts dykes and plutonic centres	constructive plate margin: opening of Atlantic
CRETACEOUS	nil		
JURASSIC	North Sea	lavas	associated with some faulting: an attempt at a constructive plate margin
TRIASSIC	nil		
PERMIAN	Midland Valley of Scotland S.W. England	basalts basalts and major granites	rift valley destructive plate margin
CARBONIFEROUS	Midland Valley of Scotland S. Pennines and S.W. England	basalts and trachytes some local volcanic activity	rift valley
DEVONIAN	S.W. England Midland Valley of Scotland Cheviot Hills Highlands and Southern Uplands of Scotland	basalts and tuffs andesites and dykes andesites and granite andesites, tuffs, major granites and gabbros	destructive plate margin: final closing of Iapetus Ocean
SILURIAN	Lake District insignificant vulcanicity		
ORDOVICIAN	Lake District North and S. Wales S.E. Ireland Southern Uplands	andesites, tuffs, granites, andesites, rhyolites, tuffs lavas, basalts	destructive plate margin: island arcs on edge of closing Iapetus Ocean
CAMBRIAN	nil		
PRECAMBRIAN	There are many outcrops of igneous rocks in the Precambrian, e.g. lavas and tuffs in Shropshire, Leicestershire and the Malvern Hills; spilitic basalts and granites in Anglesey; dolerite dykes in N.W. Scotland; peridotite in the Lizard Peninsula. It is difficult to place these into a plate tectonic framework because of their age and often deformed state, but there may have been a destructive plate margin in the late Precambrian in Anglesey.		

structive plate margin formed west of the British Isles, in a different place from the original junction formed by the closing of Iapetus. This constructive plate margin resulted in the extensive areas of Tertiary intrusive and extrusive igneous activity of Western Scotland and Northern Ireland.

Two examples of past igneous activity in Britain will be examined in more detail to illustrate the parallels they show with activity at modern plate boundaries.

Ordovician Vulcanicity

Evidence of Ordovician volcanic activity is seen principally in North Wales and the Lake District and to a lesser extent in the Southern Uplands of Scotland. The overriding characteristic of this volcanic activity is found in the andesitic lavas with a high percentage of pyroclastic rocks, which immediately suggest formation at a destructive plate margin similar to the circum-Pacific Belt of today. But is there any further confirmation of this?

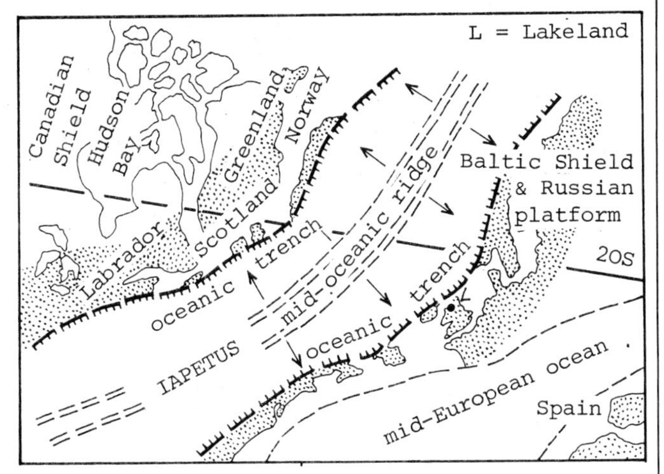

Fig. 5.1 Iapetus Ocean in the Lower Ordovician.

An ocean between England and Wales ('Europe') and Scotland and North America ('America') (Fig. 5.1) has long been suspected on palaeontological grounds. For instance, the trilobite and brachiopod fauna of England and Wales differs significantly from that found in Scotland, especially at the beginning of the Lower Palaeozoic. The Scottish fauna, on the other hand, has close affinities with that of North America. The subduction of oceanic crust at a destructive plate margin provides an answer to what happened to this ocean called Iapetus or the Proto-Atlantic.

In the Girvan-Ballantrae area of the Southern Uplands, there are spilites exposed which are thought to be a relic of this ancient ocean floor which was thrust up between the closing continents. Just where the suture between 'Europe' and 'America' lies is still a matter for some debate but it probably lies along the Solway Firth, making the Lake District the leading edge of the European plate.

Most of the evidence of volcanic activity of this period is preserved on the European side of the ancient ocean. In the Lake District, the first phase of activity is the Eycott Group. This group contains andesites and basalts of tholeiitic types which are very much like lavas found at continental margins or island arc volcanoes today. The character of the lavas indicates that they were erupted through continental crust and this has been confirmed by a recent geophysical survey of Britain. This shows there is continental crust to an average depth of 35 km beneath the area. (See the Unit *Geophysics*.)

The Eycott Group was followed by a thicker sequence of volcanic rocks, the Borrowdale Group. This group includes basaltic and andesitic lavas with a high proportion of tuffs. Some of the acidic tuffs are ignimbritic. Many of the lavas are brecciated, indicating explosive eruptions, and this feature is commonly seen in andesite volcanoes in South America today. However, the closest modern parallel to the type of activity seen in the Borrowdale Group is found in the Cascades of North America. This is part of the western Cordillera and includes Mount St Helens. Thus, a similar geotectonic situation must have existed in Ordovician times as exists in the East Pacific today, oceanic crust being subducted beneath continental crust and the consequent volcanic rocks rising through the continental crust (Fig. 5.2).

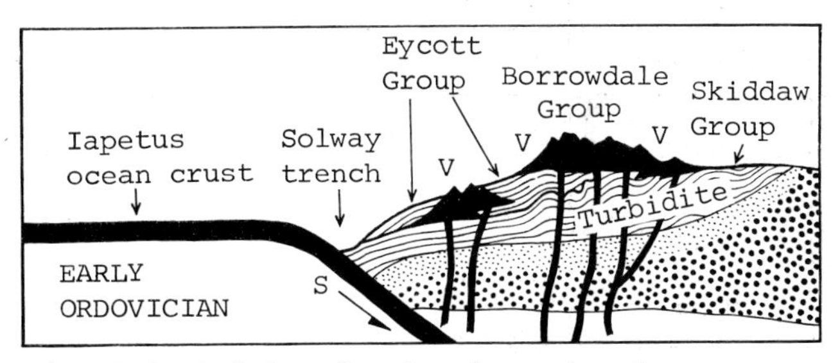

Fig. 5.2 Origin of volcanic rocks at a destructive plate margin in the Ordovician.

Similar intermediate and acidic lavas and tuffs are found in the Ordovician rocks of North Wales. There are clear indications, as indeed there are in the Lake District, that whilst some of the eruptions were subaqueous, other volcanoes built themselves up above sea level as the lavas lack a pillow structure and the ash seems to have settled out subaerially. Marine microfossils have been found in mudstones interbedded with the first tuffs which confirms that the volcanoes built themselves up through the sea rather than through lakes.

Putting all the evidence together, the Ordovician volcanic rocks of North Wales and the Lake District are clearly an ancient series of arcs underlain by continental crust and formed along the continental margin of ancient Europe.

Tertiary Vulcanicity

A contrast to the Ordovician vulcanicity is found in the Tertiary. Here the igneous activity was not related to a destructive plate margin, but rather to a constructive plate margin.

This province covers a vast area from Greenland to Iceland to Britain to Spitzbergen. Part of the province is, in fact, still active, namely central Iceland. Many of the rocks of this Tertiary Igneous Province have been referred to in previous sections. The Skaergaard Intrusion of East Greenland, the Antrim Plateau of Northern Ireland and the Ardnamurchan Ring Complex of Western Scotland are all of Tertiary age (Fig. 5.3).

The geotectonic affinities of this province are perhaps easier to appreciate than for the Ordovician as the constructive plate margin is still active in Iceland. However, a more detailed look at the Tertiary igneous rocks of Britain shows quite a complex history of intrusions and extrusions related to the opening of the Atlantic.

The initial activity in the area was the outpouring of massive amounts of basaltic lava which are seen in Antrim, and on Mull and Skye. On Mull over 2,000 m of lava are still preserved and many more metres must have been removed by both contemporaneous and recent erosion. These lava piles are made up of many individual flows varying in thickness up to 50 metres but generally 12-15 m thick. The lava flows may show columnar jointing where there has been a slow, regular cooling. The classic examples are seen at Giants' Causeway in Antrim and Fingal's Cave on Staffa, although there are many other examples.

Many of these lavas were erupted subaerially and the small amount of pyroclastic rocks shows that there was little explosive activity associated with the eruptions. This placid outpouring of basaltic lavas, many with tholeiitic affinities, is consistent with activity at constructive plate margins. The difference here is that the magma rose through continental crust to reach the surface as the two continents had not yet split apart.

In addition to the extrusive activity, there are many intrusions in the province which postdate the lavas. These intrusions are referred to as the central complexes.

Many of the central complexes comprise ring intrusions (see section 3) which are the lowest levels of the volcanoes. They were therefore intruded to a high crustal level, although many are coarse-grained. The rock types in these intrusions are both acid and basic. Ardnamurchan (Fig. 3.28) provides a fine example of a basic ring intrusion but many other igneous complexes are seen, for example on Skye, Rhum, Mull, in Northern Ireland and in the granite intrusion of Arran (Fig. 3.27).

The largest basic complex is the Cuillin Hills of Skye where the banded gabbros show a layered structure similar to the Skaergaard Intrusion. Adjacent to the Cuillin Hills lie the Red Hills formed of the acidic rock, granophyre. These are younger than the Cuillin Gabbro and it seems likely that they owe their origin to a melting of continental crustal rocks such as the Torridonian arkoses or the Lewisian Gneiss. The melting of these rocks was caused by the intrusion of large volumes of hot basic magma into the continental crust. Other acidic intrusions in the area would have had a similar genesis.

Many minor intrusions are found within the Tertiary Igneous Province and the origin of cone sheets and radial dykes has been discussed in a previous section.

However, the most revealing minor intrusions, in terms of the geotectonic origins of this province, are north-west to south-east dyke swarms (Fig. 5.3). These lie parallel to the original ridge crest and therefore to the major direction of tension within the crust at the time of the opening of the Atlantic. Their origin is therefore consistent with formation at a constructive plate margin.

The Tertiary igneous activity in Britain is confined to the Eocene, the first part of the Tertiary, and since then Britain has become inactive as the Atlantic opened up and it moved further away from the source of activity. All the igneous features observed are consistent with the initial splitting of two continents. It is interesting to note that when the Atlantic opened up it did so in a different place from the original junction of Iapetus. Scotland consequently became part of Europe rather than North America.

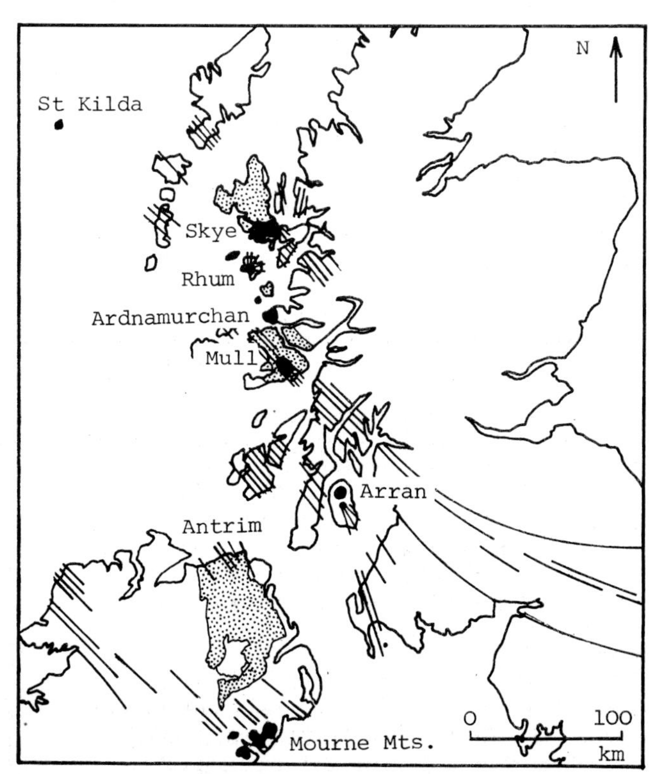

Fig. 5.3 Tertiary Igneous Province of the British Isles. Plutonic centres in black, plateau lavas dotted and dyke swarms marked by lines.